SpringerBriefs in Geography

For further volumes:
http://www.springer.com/series/10050

Xinqi Zheng · Chunlu Xue
Zhiyuan Yuan

Intensive Variable
and Its Application

 Springer

Xinqi Zheng
School of Land Science and Technology
China University of Geosciences
Beijing
China

Chunlu Xue
Zhiyuan Yuan
China University of Geosciences
Beijing
China

ISSN 2211-4165
ISSN 2211-4173 (electronic)
ISBN 978-3-642-54872-7
ISBN 978-3-642-54873-4 (eBook)
DOI 10.1007/978-3-642-54873-4
Springer Heidelberg New York Dordrecht London

Library of Congress Control Number: 2014935228

Printed on acid-free paper

Springer is part of Springer Science+Business Media (www.springer.com)

Preface

Since its introduction approximately 200 years ago, the law of diminishing returns (LDR) has had a positive impact on economic activities. However, it has been a subject to various disputes depending on the viewpoints and preconditions adopted by researchers from different backgrounds. The development of a number of new topics, including globalization, global warming, and the low-carbon movement, has brought focus to sustainable development. Thus far, sustainable development remains a concept without a generally acceptable theory. In other words, sustainable development has not been validated as extensively as other economic theories because of the lack of evaluation criteria. An investigation of sustainable development shows that the core of sustainable development is intensive development. For example, in China, conservation of water, land, energy, and materials is generally known as a push for the development of a resource-efficient, environmentally-friendly, and sustainable society. In the meantime, "red lines" for parable land, water, and forest resources have been identified as part of a national strategy for efficient, intensive land use, all of which have brought new challenges to research aimed at addressing the balance between population expansion and resource supply.

In the past few decades, few people conducted on LDR, which, given its known flaws, still provides helpful inspirations in terms of diminishing returns, marginal costs (MC), and externalities. Thus far, a sound explanation regarding the intensive margin deduced from LDR has yet to be made.

In general, despite their diversified development models, the development of countries worldwide is a process that evolves from an extensive stage to an intensive stage. Apart from in-depth integration of the LDR with intensive development, it has been impossible to summarize or explain many of these models and to provide developing and less-developed countries (particularly those with large populations and scarce land resources) with appropriate guidance and benchmarks.

Based on a comprehensive analysis and a summary of existing research results, this book proposes Intensive variables (IVs) as a theoretical foundation and the intensification function as a basic tool. Then, it combines the LDR and the intensification theory with the intensification curve. With the support of geographic information systems (GIS), the book integrates a static discussion with a dynamic analysis and data computing with spatial optimization. The result is a unique theoretical framework and methodology for the evaluation of land use efficiency, providing a generally accepted theory and methodology to guide intensive development practices. Currently, re-interpreting LDR and intensification issues plays a proactive role in the development of theories and practices in this particular field.

We believe that IV theory concerns research on the intensity of factors used in the development of human society, the substitution rule of these factors in a specific time and region, the conditional LDR and unconditional intensive development, the continuous function of intensification and the piecewise function of LDR, and the fact that diminishing returns is a special case of intensification. With regard to intensive land use, IVs may provide a scientific basis and decision-making support to the transformation of socioeconomic development models and to the optimization of their growth structures, and it could facilitate rational substitution and optimization of factors of production for higher values and benefits through integrated use.

This book uses a variety of diagrams to illustrate IVs and business cases in an interesting and logical structure as well as an easily understandable approach. In addition to the explanation of specific theories, the book also includes detailed descriptions of business cases for readers to use the theories and methodologies herein directly in their case studies and extended research.

December 2013   Xinqi Zheng

Contents

Chapter 1
Introduction

Abstract This Chapter introduces some theories supporting Intensive variable. The basic theory is the law of diminishing returns (LDR). In practice, the effect of the LDR is not distinct. Efficiency is a common goal for the use of limited land resources. Then the intensive land use concept was proposed. The concept of intensive growth was first introduced during the course of economic development. Since the introduction of the intensive land use concept, researchers globally have been trying to develop methods to determine the extent of intensive use. Therefore, we propose the "Intensive variable (IV)" as an overarching concept that covers diminishing returns, intensification, and sustainable development in an attempt to review the development of human society from a higher level with a more extensive vision.

Keywords Law of diminishing returns · Intensive land use · Intensive growth · Intensive variable

1.1 Law of Diminishing Returns

The law of diminishing returns (LDR) [1] was first proposed in 1768 by neo-classical economist Anne Robert Jacques Turgot. It argues that, when conditions for maximal product are fulfilled according to neoclassical economics theories (i.e., the increase or decrease of any factor of production results in a decline in total product), there are three typical stages:

- Stage I: Marginal product (MP) increases, and average product (AP) decreases.
- Stage II: MP declines, while AP increases.
- Stage III: AP declines.

Among other economists that have discussed this topic, the best known is that between David Ricardo and Thomas Malthus [2]:

X. Zheng et al., *Intensive Variable and Its Application*, SpringerBriefs in Geography, DOI: 10.1007/978-3-642-54873-4_1, © The Author(s) 2014

- No output exists without input. Therefore, the production line starts from the point of origin. In other words, diminishing returns do not exist in pure natural conditions without human labor.
- In the initial stage, factors of production have a maximum MP, which generally follows an "S" curve.
- Marginal returns follow a diminishing trend.

The "Iron Law of Population" proposed by Malthus in his Theory of Population used the above ideas [3]. These early economists failed to notice the effect of technological advancement on incremental production, believing that with the growing population and diminishing marginal returns, per capita production would decline continuously. However, thanks to the development of technology, production increased at a speed that outpaced the effect of population growth, resulting in continued improvement to people's living conditions [4]. The LDR shows that diminishing returns can be affected by many factors, including advances in technology, resource availability, and population quality. Early economists could not see, or foresee, variable factors. In other words, diminishing returns are phenomena observed in a closed environment in a specific timeframe.

Prior to the 19th century, LDR research in the West had been largely restricted to the agricultural sector. Short of more advanced experimentation and learning means, understanding regarding the LDR had been fragmented. Since the beginning of the 19th century, the rapid development of technologies in other sectors has paved the way for a comprehensive study of this rule from the perspectives of experimentation, math, and economics. In addition to the key condition of "keeping agricultural production technologies constant," another pre-condition, "keeping other factors of production constant," was incorporated into the LDR of land, whereby factors of production were categorized into variable and fixed factors. Meantime, the starting point of LDR research shifted from agricultural production to specific agricultural production units, leading to the introduction of the production function. By the middle of the 20th century, researchers, mathematicians, and economists had improved the LDR for land from different perspectives [5–7].

Theoretically, the reason for diminishing marginal returns is that for short-term production, an optimal balance between variable and fixed inputs exists. Marginal returns increase along with the inputs prior to the attainment of that balance, which represents maximum returns. Thereafter, marginal returns decline along with the increase of inputs [8]. In view of the entire land use process, the general law of land compensation is that the product per unit of land increases and then declines along with the increase of labor and capital inputs [9]. In general, there are three stages, as illustrated in Fig. 1.1. The X-axis shows changes in variable inputs, while the Y-axis shows the product, or compensation of land. Here, MPP means MP, APP means AP, and TPP means total product.

Sequentially, the three stages of the LDR feature insufficient, adequate, and excessive variable input, respectively, reflecting restrictions on intensive land use from an input–output perspective. Obviously, the LDR of land provides a

Fig. 1.1 Three-stage *curve* of land compensation

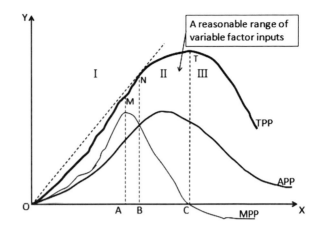

theoretical basis and a feasible approach to balancing the input and output and to achieving a scientific and intensive land use.

Land return increases or diminishes depends on the proportions of variable inputs for a certain amount of land. Product does not always increase along with input, and maximum product for land input is achieved only in Stage II. If, however, significant technological or social advances occur before returns diminish at the maximum product stage, an upward trend emerges again until the development of technology and management stabilizes, which marks the beginning of another diminishing phase [10]. Therefore, proper use of land always leads to an increase in productivity, and vice versa. The key is technology and management, which play a guiding role in the process.

In practice, the effect of the LDR is not distinct. Thanks to technological advancements, both the productivity and land returns could increase to higher levels. This, however, does not imply a non-existence of the decline of marginal returns, which is triggered by changes in the balance of the factors of production within specific technological backgrounds. In any given time and for any specific production process, a relationship between technologies, management, and all other factors that affect the input-output balance exist. Thus, there are certain limits to land as a fixed resource, the ratio between land and other variable inputs, and the acceptable amounts of such variable inputs. Beyond these limits, imbalances between land and other variable inputs would exist. As a result, incremental variable inputs would not be able to play a positive role in facilitating product and value increases. Therefore, in the course of a continued increase in inputs, the LDR of land eventually comes into play. In practice, it is necessary to take action to avoid losses resulting from the LDR. In order to maximize the product, value, benefits, and functionality of land, we must have a better understanding of the effects of the LDR [11] to be able to provide a scientific basis for intensive land use and the optimal balance between factors of production. Along with technological advancements, the industrial structure changes, as shown in Fig. 1.2. Specifically, it transforms from a labor-intensive model to a capital-intensive

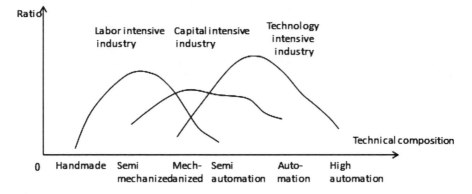

Fig. 1.2 Transformation of the industrial structure along with technology development (modified after Li and Jiapei [12]) [23]

model and then to technology-intensive and knowledge-intensive models at the higher end of the industrial chain. This is known as the general evolvement process of intensive land use.

1.2 Intensive Land Use

Regarding the use of limited land resources, efficiency is a common goal. Efficiency of land use is positively correlated with value created, and multifunctional land use has become a trend [12]. The concept of intensive land use originated from intensive agricultural production. In agrarian society, farming land was the key to mankind's survival. As man started to exploit farming land, the idea of intensive land use burgeoned. The first agricultural book in China, *Essential Techniques for the Peasantry*, published approximately 200 years ago, proposed a number of principles, such as "till the land according to your ability; it is better to till less land with high yields than to till more land with poor output [13]". These ancient proposals were a direct expression of the intensive land use concept.

In the West, the concept of intensive land use was first proposed in the 18th century by neoclassical economists David Ricard, Turgot, and James Anderson, who discovered and proved the LDR for intensive use of farming land, arguing that intensive use was the exact cause of differential land rents. With the expansion of the scope of land use and the sharpening of the conflict between population and land availability, intensive land use gradually infiltrated to urban land use. Hence, the concept of intensive use, which had been primarily focused on farmland, was introduced to the exploitation of urban land. At the end of the 19th century, neoclassical economists Ronald Coase and Yoram Barzel proposed a theory for intensive urban land use to reflect the relations between land cost and output. In the early 20th century, American land economist Richard T. Ely argued that the approach to land use by increasing labor and capital inputs was intensive land use. Since the 1950s, green belts have been established across Britain as a means to contain excessive expansion of cities, reduce the exploitation of rural land, and protect the environment. In the meantime, actions have been taken to leverage existing land and to achieve sustainability during the development process [14]. By the 1990s, the integral Smart Growth theory had emerged in the United States. In order to facilitate rational growth, it proposed a series of management tools to control and guide the use of urban land from different perspectives [15]. With the acceleration of industrialization and urbanization, the conflict between the shortage of farming land and the low-efficiency use of urban land resources in China spiked by the end of the 1990s, when studies in intensive urban land use started. In 1999, this got the attention of the Chinese government, which identified nine cities as candidates for a pilot program to assess the effects of intensive urban land use; in addition, many researchers and institutions started to select intensive urban land use as the subject of their studies. With the development of regulations, both the content and expansion of intensive urban land use has been expanding continuously [16, 17].

Intensive use is not only the premise of land-use efficiency but also a concept that has co-evolved with land-use efficiency improvements. Along with the in-depth development of research and practices in the field, the content and extension of intensive land use have been expanding continuously. Researchers [18–25] have been attempting to address the topic from different perspectives. While the arguments vary depending on specific individuals, they all agree on one point: It is necessary to increase the output of land per unit by increasing the inputs of the factors of production.

Intensive land use is a dynamic and relative concept that exists in particular regions in a specific timeframe. It is a process of continuous improvement to the efficiency and the economic, social, and ecological benefits of land by increasing input and improving management based on the premises of rational planning, optimization, and sustainability, with the goal of maximizing economies of scale and the clustering effect for local economic development. Metrics for land-use efficiency include labor, capital, technology, and material inputs in land per unit. According to the LDR, the upper limit of intensive land use is referred to as the intensive margin, whereas the lower limit is referred to as the extensive margin. It is not necessarily true that intensity is positively correlated with efficiency. In other words, maximizing the effect of intensive use should not be the sole objective in practice. Instead, actions should be taken to identify the optimal extent for the intensive growth of the land economy.

1.3 Intensive Growth

The concept of intensive growth was first introduced during the course of economic development [26]. An economic growth model is the combination of all elements necessary for the long-term development of the national economy, including land, labor, capital, technological advances, management, resource allocation, and economies of scale. In general, resources (land), labor, and capital are known as factor inputs, whereas the sum of all remaining factors is known as total factor productivity (TFP). Furthermore, economic growth models could be divided into intensive growth models and extensive growth models, depending on the importance of factor inputs and TFP to economic growth. In the extensive growth model, growth is driven primarily by incremental factor inputs. Typically, this model features substantial resource consumption, high cost, substantial obstacles to the improvement of product quality, and low efficiency. By contrast, the intensive growth model facilitates economic growth primarily through technological advances, higher efficiency of factors of production, inherent expansion and re-investment, and higher TFP, instead of incremental or detrimental physical inputs. With less resource consumption and lower cost, the intensive growth model is able to drive continued improvements to product quality and economic performance [27].

During the industrialization stage of any economy, particularly at the beginning of that stage, the extensive model (i.e., extension-oriented) is always the primary growth model, and it usually serves to consolidate the foundation of industrialization. After the industrialization stage, it typically shifts to an intensive model (i.e., content-oriented) for higher efficiency. A typical feature of the extensive growth model is the expansion of economic activities in a horizontal dimension, whereas that of the intensive model is in-depth development. The development of any economy must transition from extensive, extension-oriented expansion to intensive, content-oriented growth, that is, from a low-end model to a high-end one. This is a general law of socio-economic development.

In order to achieve and continuously move toward intensive, content-oriented growth, actions should be taken to (1) improve the efficiency of productive resources; (2) ensure that efficiency improvement outperforms resource input; and (3) continuously increase the weight of efficiency improvement in economic and intensive growth to drive the extent of intensification from the less intensive, moderately intensive to highly intensive.

Thus, determining the position of the land economy in the transition from extensive to intensive growth, content-oriented growth is of interest. In real economic activities, the extensive and intensive models exist and interact with one another. Extensive growth is the foundation and premise of intensive growth. By contrast, intensive growth is the extension and evolvement of extensive growth. While technological advancement-driven intensive growth is an inevitable result of extensive growth, no intensive growth can be completely free of extensive elements. In modern economics, no extensive growth depends completely on incremental factor inputs and zero technological advancement; furthermore, no intensive growth exists without incremental factor inputs. In general, however, human society uses intensive growth.

1.4 The IV Proposal

Since the introduction of the intensive land use concept, researchers globally have been trying to develop methods to determine the extent of intensive use. Prior to the 1930s, land evaluation methods had been developed in the United States, Russia, and Germany for the purpose of taxation. The most typical models include the Storie Index Rating (SRT) and the Cornell System, which were introduced in the United States in 1933; the Arable Land Evaluation Rules issued by the French Ministry of Finance in 1934; and the Land Index Rating System issued in Germany in the 1930s. In the United States, the first comprehensive land evaluation system was introduced in 1961 based on the Land Capability System (LCS), which was initially developed in the 1930s. In 1981, the US Department of Agriculture (USDA) adopted the Land Evaluation and Site Assessment (LESA) strategy guide for the evaluation of federal land. In the past two decades, more achievements have been made in urban land evaluation research. Many evaluation models have been developed, including the general ecosystem model (GEM), the Patuxent landscape model (PLM), the conversion of land use and its effects (CLUE), the spatial Markov model, and the land use change analysis system (LUCAS) [28]. In China, the entropy method, the fuzzy theory-based evaluation method, the main ingredient analysis method, the analytic hierarchy process (AHP), the Delphi method, the integrated multi-factor evaluation method, the BP network model, the RS-based artificial neural networks (ANN) model, and the Grey relational analysis evaluation method have been adopted to evaluate the extent of intensive urban land use. Although different methods apply to different subjects and scopes, each has its advantages and limits, as shown in Table 1.1.

Each of these methods has its advantages and disadvantages while focusing on one particular aspect of intensive land use. However, none of them attempts to address the issue of intensive use from the perspective of macroscopic content of intensification, nor do they provide a time-spatial calculation. From the macroscopic content perspective, intensive land use is the result of the exploitation of natural resources by human society. From the time-spatial viewpoint, economic development varies over regions and periods, as does intensive land use. In other words, the efficiency of resource utilization varies in regions and periods. Since the introduction of the LDR, disputes have focused on its basic concept and content, on whether the LDR exists, and on a few LDR-related qualitative judgments; furthermore, few studies have conducted thorough investigations on the topic. In addition to global population growth, the world is increasingly running out of resources or reaching the limits of its capacity. To this end, we believe that intensive growth is the pursuit of mankind and a general goal of sustainable development for human society. In other words, the pursuit of sustainable development is essentially one for intensive growth; furthermore, diminishing returns is an inherent aspect of socio-economic development or a phenomenon observed in a closed environment in a particular period during the course of intensive growth. Therefore, we propose the "Intensive variable (IV)" as an

Table 1.1 Urban land use evaluation models

Model	Advantages	Disadvantages
General ecosystem model (GEM; dynamic)	Time span adjustable according to specific evaluation target; adding/deletion of evaluation factors allowed	Contains less impact factors
Patuxent landscape model (PLM; dynamic)	Builds on the advantages of GEM while incorporating additional variables; suitable for evaluating the impact of land management and optimal management practice	Contains less impact factors
Conversion of land use and its effects (CLUE; disperse, limited)	Extensive evaluation factors with large time and spatial spans	Less consideration for systemic and economic factors
Analytic hierarchy process (AHP)	Provides quantitative metrics to establish an intensive land use metrics system	Improper selection of factors could result in confusing definitions; furthermore, inadequate relations between factors would impair the results
Delphi method	Allows collection of opinions through multiple rounds of anonymous feedback from experts and data aggregation	The selection of experts proves to be quite challenging; the results of evaluation could be subjective
Grey relational analysis evaluation method	Less demanding on sample size; no typical sample distribution rule required for analysis	Defective calculation methodology
Main ingredient analysis method	Uses relevant statistical theories to address the correlation between metrics; weights are identified in accordance with specific amounts of contribution, and are therefore objective	Assumes multiple linear correlation between the metrics, limiting the scope of application when the metrics are non-linearly correlated
BP network model	Satisfactory result could be expected in practice	The value assigned to each controlled part is largely subjective; the impact of the controlled parts is hardly considered
Integrated multi-factor evaluation method	Simple, easy to understand, capable of reflecting the overall utilization of land	Integrated representation and metrics not unified; weights determined largely on a subjective basis

(continued)

Table 1.1 (continued)

Model	Advantages	Disadvantages
Fuzzy theory-based evaluation method	Scientific, reasonable, practical, quantitative evaluation, with abundant information	There's yet to be a well-established, effective Membership Function with less complicated calculation
Entropy method	Capable of precisely reflecting the practical value of the metrics and determining their weights. It is an objective weight assignment method that produces reliable, accurate data	Does not provide inter-data comparison; weights vary along with the metrics

overarching concept that covers diminishing returns, intensification, and sustainable development in an attempt to review the development of human society from a higher level with a more extensive vision.

Without a set of generally accepted criteria for comparison among countries, a global awkward exists whereby countries take various positions on global issues, for example, in defining developed, developing, and less developed countries. We believe that this definition should consider not only GDP and per capita GDP but also the quality and structure of growth. We expect that the IV is able to reflect not only growth quality and substitutability among various factors but also the development structure.

References

1. Shephard, Ronald W. 1970. Proof of the law of diminishing returns. *Zeitschrift für Nationalökonomie* 30(1–2): 7–34.
2. Ricardo David. 1815. An essay on profits. Extracts in Mogens Boserup.
3. Malthus Thomas Robert. 2013. *An essay on the principle of population.* 1. Cosimo, Inc.
4. Encyclopedia Britannica. 2012. *Diminishing returns.* Encyclopedia Britannica, Inc. 16 Feb 2012.
5. Shephard, Ronald W. 1969. *Proof of the law of diminishing returns.* Berkeley: Department of Industrial Engineering and Operations Research University of California.
6. Shephard, Ronald W., and Färe Rolf. 1974. *The law of diminishing returns. Zeitschrift für Nationalökonomie.* Berkeley, Calif., USA, Lund, Sweden, 34, 69–90.
7. Fogarty, G.J., and L. Stankov. 1995. Challenging the law of diminishing returns. *Intelligence* 21(2): 157–174.
8. Gao, Hongye. 2000. *Western economics.* Beijing: China Renmin University Press.
9. Yu Chen. 2012. *Research on potential exploitation and policy of urban land intensive use: a case of Wuhan City.* PhD thesis. Wuhan: Huazhong Agricultural University.
10. Haiyan Chen. 2011. *The mechanism of land intensive use change under the background of economic development.* PhD thesis. Nanjing: Nanjing Agricultural University.
11. Zhihong Zhang. 2006. *Research on modelling the intensive use of urban land.* Master's thesis. Nanjing: Nanjing Normal University.

12. Li, Yaoxin, and Wu, Jiapei. 1992. The change of technological component in the evolution of industrial structure. *The Journal of Quantitative and Technical Economics* 9(11): 54–59.
13. Kong, Fanwen, and Xu, Yumei. 2010. The historical trajectory of intensive land use. *China Land* 5: 53.
14. Ma, Yi. 2003. Introduction and learn from English land management system. *China Land* 12: 39–40.
15. Chen, Shuang. 2006. The enlightenment of American policy to promote the intensive use of land for construction. *Journal of Hubei University (Philosophy and Social Science)* 6(33): 741–745.
16. Zheng, Xinqi. 2004. *Industrial optimize the allocation of intensive use evaluation of urban land-theory, methods techniques and empirical.* Beijing: Science Press.
17. Huijuan Si. 2007. *Study on urban land use intensive thesis and evaluation method—a case of Qinghai province typical cities.* Master's thesis. Xi'an, China: Chang'an University.
18. Baode, Bi. 1991. *Land economics.* Beijing: China Renmin University Press.
19. Yue, Jinfeng. 1992. *Real estate law dictionary.* Pretoria: Law Press.
20. Tao, Zhihong. 2000. Several basic problems on the city land intensive use. *China Land Science* 14(5): 1–5.
21. Zhao, Pengjun, and Jian Peng. 2001. High efficient and intensified use of urban land and its evaluation index system. *Resources Science* 23(5): 23–27.
22. Lin, Jian, Qizhen Qi, and Jingyao Jin. 2004. How to use land—the connotation and indicators evaluation of urban land intensive use. *China Land* 9: 4–7.
23. Xie, Min, Jinmin Hao, Zhongyi Ding, et al. 2006. Connotation and evaluation index system of urban land intensive use. *Journal of China Agricultural University* 11(5): 117–120.
24. Yang, Shuhai. 2007. Connotation and evaluation index system of urban land intensive use. *Inquiry into Economic Issues* 1: 27–30.
25. Wang, Jiating, and Kaiwen Ji. 2008. Study on dynamic mechanism of urban land intensive use. *Urban Problems* 8: 9–39.
26. Shilong Xu. 1997. Defining connotation and quantitative calculation of intensive economic growth. *Statistics and Information Tribune* 6(2): 6–13.
27. Mei, Zhou. 1993. The progress of science and technology and intensive management. *Social Sciences in Hubei* 1: 8–10, 25.
28. Yu, Zhimin, and Zhiqiang Yan. 2012. Research review on common evaluation method of urban-land intensive utilization assessment. *Urbanism and Architecture* 17: 252–253.

Chapter 2
The IV Theory

Abstract The IV theory is a new theory that focuses on the quality and intensity of socioeconomic development and the substitutability and structure of factors. With the IV theory, we are now able to abstract complex social activities into simple mathematical functions. We set up the models of IV theory based on the definition, and explain the hypotheses of the theory. Then the IV functions are created on two different scenarios of resource conditions. In the end of this chapter, we obtain the IV curve based on the LDR diagram.

Keywords Intensive Variable theory · IV models · IV curve

2.1 IV

People receive benefits, value, or utility through exploitation of natural resources. In other words, natural resources would not have any value for human society without human involvement. Furthermore, resources first became valuable to mankind after the emergence of human society, when people started to "process" natural resources using labor; thus, LDR is similar. Human society, from its primitive versions of fishing and hunting to the agrarian and then the industrialized societies, has been continuously evolving from extensive to intensive natural resource use. However, it was not until the 19th century that the concept of intensive use was invented. Without human exploitation, natural resources would not have socioeconomic value. In other words, the socioeconomic value or utility of natural resources are realized only through the production activities of people, and human activities have added value to natural resources. Having realized that extensive use would only exhaust the natural resources available, people eventually adopted intensive use—a concept invented based on consumption of natural resources by human society. Obviously, intensive use is closely related to human activities and natural resources.

X. Zheng et al., *Intensive Variable and Its Application*, SpringerBriefs in Geography, 11
DOI: 10.1007/978-3-642-54873-4_2, © The Author(s) 2014

The word "intensive" means involving great effort or work and enhanced farming in agriculture. In other words, an intensive process is one with steady enhancement, and IV is a variable that reflect continued incremental changes or the process of intensification.

By definition, an IV is a variable that results in an increase of value, benefits, or utility of natural resources through human activities. It is expressed in the value, benefits, or utility per unit of natural resources. IV is the capable of reflecting the status or extent of intensification of a particular region in a particular period. The IV theory is a new theory that focuses on the quality and intensity of socioeconomic development and the substitutability and structure of factors.

With the IV theory, we are now able to abstract complex social activities into simple mathematical functions. Then, we can summarize the general laws of social production by analyzing and studying the attributes of these functions. We can predict the development trends of specific regions by modeling future intensive use to support socioeconomic development and resource planning.

2.2 Models of IV Theory

As previously mentioned, IV is based on activities related to natural resources. The relationships between human activities, and natural resources can be expressed using the following function:

$$I = f(R, H) \tag{2.1}$$

where I is the IV, R represents resource conditions, and H indicates human activities. Natural resources have always been available on Earth, but human activities appeared significantly later. Without human activities, natural resources would not have any value or benefit, and the concept of IV would not exist. In other words, if $H = 0$ and $R = r$ (where r is a constant), $I = 0$. As mankind started to exploit natural resources, they produced value, benefits, and utility, and hence, IV appeared. Therefore, IV can be expressed as a function with human activities and natural resource conditions:

$$I = R \times H \tag{2.2}$$

Human activities can be further detailed. In general, human activities include inputs that can help increase the value, benefits, and utility of natural resources, including capital input (K), labor input (L) and overall technological advancement (T). They can be expressed using the following equation:

$$H = K \times L \times e^{T} \tag{2.3}$$

where the overall technological advancement is an enhancer. IV exists with or without technology, which serves only to provide additional improvements. Therefore, $T \geq 0$. Similarly, without additional cost, optimization or resource allocation through better management could also help increase resource value, benefits, and utility. If this factor is considered, it overshadows the intensity of overall technology advancement. Management improvements could help optimize the combination of labor, capital, and equipment, which, while relevant to technology, would be helpful to list separately given the actual situation. Therefore, optimized management (M) is included in Eq. (2.3), obtaining:

$$H = K \times L \times e^{(T+M)} \tag{2.4}$$

Then, the elasticity coefficient is incorporated and additional adjustments are made to change the function of the IV theory as follows:

$$I = R \times L^{\alpha} \times K^{\beta} \times e^{(T+M)} + \delta \tag{2.5}$$

where α and β are elasticity coefficients of capital input (K) and labor input (L), respectively, and $0 \leq \alpha \leq 1$ and $0 \leq \beta \leq 1$. No elasticity coefficient is identified for resource conditions (R), an index variable, and δ is the error coefficient.

Set $\mu = e^{(T+M)}$, where the function resembles a Cobb-Douglas (C-D) production function proposed by C. W. Cobb and P. H. Douglas in 1928 in an attempt to address the relations between input and output [1, 2]. The C-D production function improves the general production function and holds an important position in theoretical research and practice in quantitative economics or econometrics [3, 4]. Holding other conditions constant, its mathematical model is as follows:

$$Y = A_0 L^{\alpha} K^{\beta} \mu \tag{2.6}$$

where Y is total yield, A_0 is overall technology level of year 0, L is labor input, K is capital input, α is the elasticity coefficient of labor output, β is the elasticity coefficient of capital output, μ represents random interference, and $\mu \geq 0$.

The C-D production function assumes that changes to total yield are the result of changes only to labor and capital, and that labor and capital are used at the same intensities each year. Therefore, cyclical fluctuations in yields are not considered in this function [5]. There are flaws in the C-D production function, including the following: (1) Production is the result of scale expansion, without considering the effect of technological advancement. In addition, knowledge, skills, and experience of workers and managers are expected to improve, and the use of plant houses, equipment, capital, and technologies is expected to become increasingly rational; thus, total yield changes over time given that advancements in technology vary over periods. (2) It does not consider the effect of natural conditions or social and political factors on production activities. Many improvements have been proposed to address this issue, for example, by incorporating the missing factors into the function. Many production functions are based on the C-D function [6, 7], such as the constant elasticity of substitution (CES) production function developed by Arrow et al. [8]. This function appears to be more complex but is theoretically better than the C-D production function, described as follows:

$$y = A[\alpha L^\rho + \beta K^\rho]^{1/\rho} \tag{2.7}$$

where α is the labor distribution ratio indicating the labor intensity of a specific technology; β is capital distribution ratio indicating the capital intensity of a specific technology; $\alpha + \beta = 1$; and ρ is the elasticity of substitution.

Taking into consideration changes to the elasticity of substitution, all other production functions can be regarded as variations of a CES production function, which therefore has a much wider scope of applications [9]. Furthermore, it is possible to prove that the C-D production function is a special case of the CES production function [10]. For CES production functions, the elasticity of substitution for labor α and capital β are correlated with the technology coefficient K/L as follows:

$$L \cdot K = \left[\frac{\beta P_K}{\alpha P_L}\right]^{1/(1+\rho)} = \left[\frac{\beta P_K}{\alpha P_L}\right]^\sigma \tag{2.8}$$

The results of CES calculations are more accurate than those of the variable elasticity of substitution (VES) and therefore better reflect the size of the elasticity [11, 12]. In addition, the production function changes along with technological advancements. In order to more accurately reflect the substitutability of the factors, Revanka proposed the VES production function in 1971. Both CES and VES are monotonic functions. To provide an elasticity of substitution that is a non-monotonic function, Fare and Yoon argued the WDI production function. However, measuring the parameters of the WDI production function proves to be an obstacle to its application [13].

The production function theories show that in the process of using land to produce products and services necessary for human society, certain relations of substitutability between land, capital, and labor exist. By increasing capital and labor inputs, less land could be consumed for given products to satisfy the growing physical and cultural demands of human society. People have been using labor and capital as substitutes of land; in other words, we have been trying to increase total product per unit of land to satisfy the growing physical and cultural demands of human society by increasing the variable inputs. Essentially, this is a process of intensive land use.

As a matter of fact, intensification is the increase of intensity through a number of approaches, such as improving technology and management, increasing labor and capital inputs, and optimizing structures and spatial distributions. These factors are inter-substitutable, meaning that the substitution of one with another could help achieve intensive use.

With statistics-based regression analysis, we can obtain the optimal labor input (L^*) and optimal capital input (K^*) per unit of land [14]:

$$L^* = \sqrt[\alpha+\beta-1]{\alpha^{\beta-1}K^{\beta}/\beta^{\beta}AL^{\beta-1}Y} \tag{2.9}$$

$$K^* = \frac{\alpha L}{\beta K}L^* \tag{2.10}$$

The extent of intensive land use can be reflected with intensive land use coefficient E:

$$E = \frac{AL_0^{\alpha}K_0^{\beta}}{AL^{*\alpha}K^{*\beta}} = \left(\frac{L_0}{L^*}\right)^{\alpha}\left(\frac{K_0}{L^*}\right)^{\beta} \times 100\,\% \tag{2.11}$$

where K_0 and L_0 represent capital and labor inputs per unit of land with a given purpose. E is positively correlated with capital and labor inputs and, hence, the extent of intensive land use [15, 16].

In summary, we have identified the following identities for the IV theory:

(1) Because the IV theory is concerned with the actual intensity of resource use, the IV should be larger than 0.
(2) The IV is a variable with a changing value depending on period and location.
(3) As IV may arise from the interaction between human activities with different resources, it has a wide range of forms. In other words, IV has different meanings and forms for each resource and economic activity.
(4) The IV calculation may entail as many samples as possible, because it needs to apply probability as well as statistics knowledge.
(5) IV models are just abstract summarizations of extremely complex human activities. In practice, different conditions should be added to make sure that these conditions fit the reality of human activities. In the meantime, it is necessary to keep these conditions in mind when analyzing the intensification of human activities.
(6) The IV models reflect the LDR. Therefore, an IV reflects a continuous dynamic instead of an unchanging process of human activities.
(7) Each factor of the IV model can be divided infinitely, allowing for the use of infinitesimal analysis.

2.3 IV Hypotheses

While development and exploitation of natural resources generate economic benefits, excessive development has resulted in excessive consumption of resources, which, in turn, has caused a decline of product or yield and irreversible degradation of the ecological environment. Thus, the ideas of conservation and intensive use of natural resources have been proposed. The IV theory was developed in the background of mankind's exploitation of natural resources. In other words, IV would not exist without the development and exploitation of natural resources. Exploitation results in consumption or quality degradation of natural resources. With technology, however, people restore their ability to generate benefits. Therefore, the hypotheses of IV theory include:

(1) There are natural resources available, that is, $R > 0$;
(2) There are human activities affecting natural resources and generating benefits, value, and utility, that is, $I > 0$;
(3) There are changes to technologies that help continuously generate benefits, value, and utility from natural resources, that is, $T \geq 0$;
(4) There are changes to management ability that help continuously generate benefits, value, and utility from natural resources, that is, $M \geq 0$.

2.4 Intensive IV Functions

IV means the value, benefits, or utility per unit of natural resources. Using Q as the value, benefits, and utility of total product and S as the quantity of resources, the IV function can be expressed as follows:

$$I = \frac{Q}{S} \tag{2.12}$$

In line with the function models of IV theory, we focus our discussions on two different scenarios of resource conditions.

(1) Fixed Resource Conditions

The effect of fixed resource conditions on the IV in different stages remains the same. In this scenario, the value of R is 1, and the IV function can be simplified as follows:

$$I = L^{\alpha} \times K^{\beta} \times e^{(T+M)} + \delta \tag{2.13}$$

For example, consider a case study of intensive use in a particular region. If the total volume of resources is kept constant and no significant change occur to resource quality and other attributes, this function can be used to measure the extent of intensive use.

(2) Variable Resource Conditions

Variable resource conditions have different effects on the IV in different stages. In this scenario, different values are assigned to R for calculation. Because it is impossible to obtain quantitative resource conditions, assessments are made on the suitability of the resources, and the result is used as the value for R. Then, the IV function changes to the following equation:

$$I = RL^\alpha K^\beta e^{(T+M)} + \delta \tag{2.14}$$

Function (2.14) is referred to as the basic function of IV.

Generally, resource conditions of any region are variable. Therefore, to simulate the IV function over different periods, we can use the following:

$$I = R(t)L^\alpha K^\beta e^{(T(t)+M(t))} + \delta \tag{2.15}$$

or

$$I = R(t)e^{(T(t)+M(t))}(L^\alpha \times K^\beta)^{1/\rho} + \delta \tag{2.16}$$

In the event that more factor inputs are required, the following function is acceptable:

$$I = R(t)e^{(T(t)+M(t))}(\alpha_1 K^\rho + \alpha_2 L^\rho + \cdots + \alpha_n M^\rho)^{1/\rho} + \delta \tag{2.17}$$

Variables in these IV models should have different values depending on specific targets of research and purposes of resource utilization. Instead of fixed values, they should have variable values depending on the actual situation and the purpose of the study. For example, a variable could be the economic value created per unit of construction land, the agricultural product generated per unit of farming land, or the aquatic breeding output per unit of water area. Resource conditions could imply any one or more land, air, water, and mineral resources. Capital and labor input are mutually substitutable depending on the specific target of evaluation. For example, metrics used for an intensive land use evaluation of one particular development zone in China could include land use status, benefits, and management performance. In this case, management performance is integrated into T, whereas land use status and benefits are categorized into capital input (K) and labor input (L), and the function is changed into a model better suited to intensive land use evaluation for development zones in China [17].

2.5 IV Curve

Given the limited nature of land resources, our goal is to maximize the efficiency of land use, which can be achieved with one premise—intensive use is usually connected to high performance. Intensive land use is a dynamic, relative concept specific to a particular region and a particular time, and it is the process of

Fig. 2.1 LDR under
changing technology
conditions

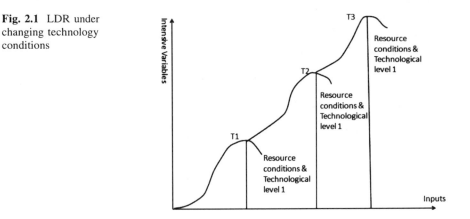

continuously improving the efficiency of land use to obtain satisfactory economic, social, and ecological benefits through rational planning, optimization and sustainable growth, additional inputs, and better management while meeting the requirements of regional development and maximizing economies of scale and the clustering effect under current conditions. The intensities of labor, capital, and technology inputs per unit of land are metrics to measure the extent of intensive land use. Factors that affect the efficiency of land use include land supply, cost, usage and location, technology, market capacity, land-related speculation, taxation, policies, and systems. Higher intensity does not necessarily mean higher efficiency. In reality, maximal intensity is not an exclusive goal, and an adequate intensity is a better objective for ensuring maximization of land use efficiency.

One of the main theoretical bases of intensive use is the LDR, meaning that at the point of diminishing returns, additional inputs to land would result in zero MP and maximal total product. Intensity of land use reaches its peak at this critical point, which is also known as the intensity margin of land use. Throughout its history, mankind has been progressing along a journey toward this critical point. Today, LDR has become a shadow law that exists in our mind, even though in reality, it is non-existent.

LDR is a component of the IV theory, implying here that with given production technologies and all other factor inputs kept constant, the additional input of a single factor would result in increasingly less incremental product after a certain point. According to the IV theory, resource conditions and technologies are constantly changing, and the balance and substitutability among factors of production change as well. As resource conditions and technologies change, marginal returns of inputs change accordingly, as shown in Fig. 2.1.

With resource conditions and technologies kept constant, an increase of a single input would result in the same total product, MP, and LDR curves. When resource conditions and technologies change, total product peaks and continues to increase, while MP declines to a negative level before increasing again. In reality, as

Fig. 2.2 Development stages
of IV

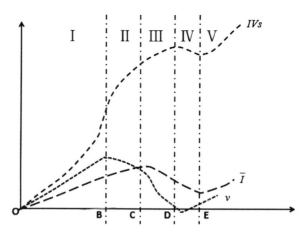

advances in human civilization have been driving the evolvement of technologies
and resource conditions, the MP curve would bottom out before it actually drops to
zero.

Based on this LDR diagram, we can obtain the IV curve as shown in Fig. 2.2.
Five stages for IV exist depending on the developments of MP v and AP $\bar{I}$.

(I) v increases (before reaching its peak) and $\bar{I}$ increases; this is a stage of low
intensity of use;

(II) v declines and $\bar{I}$ continues to increase (before reaching its peak); this is a
stage of medium intensity of use;

(III) v declines (before dropping to zero) and $\bar{I}$ declines; this is a stage of intensive
use;

(IV) v is negative; regardless of the state of $\bar{I}$, this is a stage of excessively
intensive use;

(V) v bottoms out and rises above zero and $\bar{I}$ increases; this is a stage of post-
intensive use, after which another round of use would begin under new
conditions (as shown in Fig. 2.2).

According to these classifications, the elasticity coefficient is obtained through
fitting. Adjustments are made to keep technologies and other conditions constant,
while capital and labor inputs increase separately to enable identification of specific stages and analysis of the status of intensive land use.

The IV curve integrates LDR and intensification into a single model, expanding
static LDR to dynamic contexts and driving the development of the original theory. IV applies to all land use cases, and represents the overall land use situation
within the society.

References

1. Tian, HuoMing. 1994. *The production function and agricultural economic analysis.* Chengdu: Southwestern University of Finance and Economics Press.
2. Liu, Zhiting, and Zhang Min. 2006. Calculational analysis on the contribution of science and technology progress to the economic growth in Qingdao. *Journal of Qingdao University of Science and Technology (Social Sciences)* 22(4): 49–55.
3. Dong, Xiaohua, Xin Wang, and Li Chen. 2008. Review of Cobb—Douglas production function theory. *Productivity Research* 3: 148–150.
4. Yuan Wang. 2005. *The evaluation of urban infrastructure investment efficiency based on Malmquist-DEA index.* Master's thesis. Tianjin: Tianjin University.
5. Jun Zhang. 2002. *Estimated production theory and production function.* Master's thesis. Beijing: Graduate School of China Academy of Social Sciences.
6. Li, Zhaoping. 1997. Comparisons of CD production function with CES production function. *Operations Research and Management Science* 6(3): 101–103.
7. Zhao, Linan, and Yanan Zhao. 1994. The calculation model and empirical analysis of technological progress and returns to scale. *Quantitative and Technical Economics* 8: 43–47.
8. Arrow, K.J., H.B. Chenery, B.S. Minhas, et al. 1961. Capital-labor substitution and economic efficiency. *The Review of Economics and Statistics* 43(3): 225–250.
9. Xiao, Donghua, and Huiyuan Yao. 2009. Study on the return of migrant workers problems based on CES production function model. *Lanzhou Academic Journal* 8: 87–91.
10. Wang, Weinan, and Wenju Wang. 2011. The balanced analysis of optimal allocation of three industrial based on CES production function. *Journal of Jilin Normal University (Humanities and Social Science Edition)* 39(6): 44–46.
11. Xiaoyun Jin. 2008. *An empirical study on the alternative relationship between capital and land.* Master's thesis. Zhejiang: Zhejiang University.
12. Cheng, Maolin. 2013. Correction and empirical analysis of the CES production function model. *Chinese Journal of Engineering Mathematics* 30(4): 535–544.
13. Jiayun, Xu, and Shuyun Li. 2012. Study on the overseas talents backflow based on CES production function. *Forum on Science and Technology in China* 12: 102–106.
14. Tangqi Xu, Dengping Ju, Pingping Xu, et al. 2009. *Study on the land intensive use evaluation method based on the production function theory.* Proceedings to 2009 Annual Conference of China Land Science Society 30 Nov 2009.
15. Zhu, Yanguang. 1984. Some views on the application rationality of Cobb Douglas production function. *Economic Science* 6: 12.
16. Solow, K.M. 1957. Technical change and the aggregate production. Review of Economics and Statistics.
17. TD/T 1030-2010. *Standard for evaluation of land intensive use of development zone.*

Chapter 3
IV Methodology

Abstract This chapter mainly introduces IV methodology. Parameter estimation of IV method is used as a reference for the parameter estimation of IV functions, which can then use relevant statistics on IV, capital and labor inputs. The resource suitability is estimated by resource condition suitability assessment, and technological advance intensity is calculated by Malmquist Index. At last, we describe the calculation of resource structure and distribution optimization parameter, which include distribution optimization parameter and time change parameter.

Keywords Parameter estimation · Malmquist index · Resource suitability · Gini coefficient

3.1 Parameter Estimation for IV Function

Identifying parameters for IV functions is a challenge. We drew from the experience of similar models, such as the linear CES method proposed by Kmenta and tested by Kadane, and found that this model is fairly reliable for an efficiency parameter estimation but less reliable for an elasticity estimation. In his non-linear Bayesian regression model, Lancaster listed a number of parameter estimation methods for the CES function. Berndt and Antràs used CES optimization conditions to build multiple equations for parameter estimation and concluded that in the United States, industrial product generally fits the C–D production function [1–3]. In his doctorial essay, Xu Zhoushun adopted the Bayesian model and conducted a parameter estimation based on the CES function. He found that using small sample sizes for Bayesian model-based CES parameter estimations would address some of the flaws resulting from insufficient data and obtain valid estimates [4].

Because points beyond the content reflected by the model are not our concern and we want to investigate only the model itself, Parameter Estimation of IV method can be used as a reference for the parameter estimation of IV functions, which can then use relevant statistics on IV and capital and labor inputs.

X. Zheng et al., *Intensive Variable and Its Application*, SpringerBriefs in Geography, DOI: 10.1007/978-3-642-54873-4_3, © The Author(s) 2014

By contrast, resource conditions are indicators that must be considered from multiple perspectives. They are the result of interaction between multiple factors [5], including location, quality, accessibility, and demand for resources. An integrated, multi-factor evaluation is a method whereby a number of factors are used for the assessment or quantization of the resources. The underlying idea is to convert multiple indicators into an integrated evaluation indicator [6]. Typically, a number of methods, including AHP, fuzzy analysis, GRA, DEA, BP, and Delphi analysis can be used for an integrated assessment of these factors. In certain circumstances, two or more of these methods are used simultaneously. Land suitability assessment is a type of multi-factor evaluation, and its purpose is to determine whether and how a land plot is suitable for a particular use. Land suitability assessment is the basis for sound land use decision making and planning. Its underlying theory is that given the existing productivity and land use method, assessments are made on the suitability of the land plot to a particular use and its quality and known restrictions using natural and social attributes of the land plot as indicators for the assessment to enable classification and rating of the land plot. Based on a comparison of these multi-factor evaluation methods, theories, and contents of IV, we found that land suitability assessment is a better option. Therefore, it is necessary to conduct a land suitability assessment on the resource conditions and use the result as an indicator.

The intensity of technological advancement is an integrated indicator that reflects technological developments within a given period. Typically, the DEA, GRA, and fuzzy models are used to evaluate the level or contribution of technologies. An indicator that has been extensively accepted in recent years is the Malmquist index (MI) developed by Malmquist in 1953. In 1982, Caves, Christensen, and Diewert used the index to estimate changes to production efficiency. In 1994, Rolf Färe et al. combined a linear, non-parameter planning method of this theory with the DEA theory, greatly expanding the application scope of the MI [7]. Therefore, it is feasible to use the MI to estimate the intensity of technological advancements and indicate the extent of technology development in the IV function, a suitable candidate for the index.

Based on abundant available data, we used statistical models to develop the following regression function between IV and other variables.

Set (2.14) to $e^{(T+M)} = T$, ignore the error term, and use natural logarithms on both sides to obtain the following:

$$\ln I = \ln R + \alpha \ln K + \beta \ln L + \ln T \qquad (3.1)$$

Accordingly, a growth equation can be developed for the model:

$$\frac{\Delta I}{I} = \frac{\Delta R}{R} + \alpha \frac{\Delta K}{K} + \beta \frac{\Delta L}{L} + \frac{\Delta T}{T} \qquad (3.2)$$

where $\frac{\Delta I}{I}$ is the growth rate of the IV; $\frac{\Delta R}{R}$ is the change rate at which resource conditions are compared with technological advancements; $\frac{\Delta K}{K}$ is the growth rate of

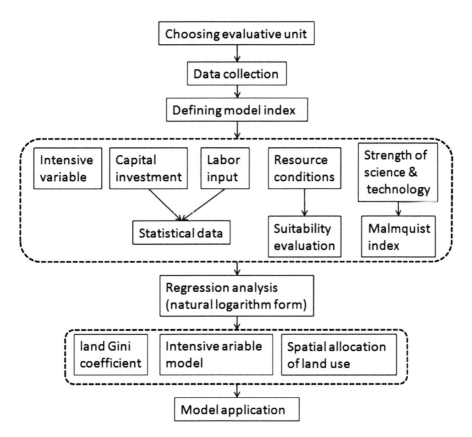

Fig. 3.1 Parameter estimation processes of IV function

capital input; $\frac{\Delta L}{L}$ is the growth rate of labor input; and $\frac{\Delta T}{T}$ is rate of technological advancement.

Hence, we can get the elasticity coefficients α and β, the elasticity coefficients for capital and labor inputs, and the corresponding IV function. Considering the attributes of rationality of land use structure and its spatial distribution, we can use the Gini coefficient to analyze the rationality of the land use structure [8]. Furthermore, we can use the space optimization model to analyze the spatial identities of IV [9]. In practice, considering actual land use situations in rural and urban areas, different values should be selected for the variables to enable model fitting. Various models can be developed for the land use situations of different regions, such as East Northern China and Western China. It is also possible to develop different indicators based on regional cultural backgrounds to enable correlation with the IV function model.

See Fig. 3.1 for the parameter estimation processes of the IV.

Many methods described in the literature regarding similar production functions can be used as references for parameter estimation for the IV function. However, in view of the new resource conditions and overall technological advances specified in this study, a better method is required for parameter estimation.

3.2 Resource Condition Suitability Assessment

Resource suitability refers to the suitability of a specific resource type to certain uses on a continued basis, including whether and how suitable it is. The term is often used to describe the availability of particular resource conditions for intensive use. In general, suitability assessment is the process of determining the most suitable resource exploitation method in line with specific requirements [10, 11]. Because of various understandings regarding the content and applications of resource exploitation, there are a number of different suitability assessment methods. Studies on IV using land resources as an example are discussed in the last sections of this book. Here, we would like to focus our discussion on land suitability assessment. As different land resource conditions are required for different land types, land suitability assessment is meaningful only for land of designated types.

Significant differences have been observed in land suitability, and certain land types have very extensive suitability. For example, farmland is suitable for agriculture, forestry, and husbandry purposes, and it is known as multi-purpose suitability. Some land types are suitable for only one or two purposes, that is, single-purpose or dual-purpose suitability [12]. Land suitability assessment is the process of determining whether a plot of land is suitable for a specific purpose and how suitable it is [13]. Essentially, it is a means for handling the relations between land quality and land exploitation (including the land quality requirements of each exploitation method).

Land suitability assessment can be performed through spatial and non-spatial methods. Spatial methods are based primarily on GIS and vector-raster mixed data models. Non-spatial methods are used to conduct assessments mainly through the collection of relevant data and the selection of adequate models. Currently, mainstream land suitability assessment methods include overlay analysis [14, 15], the multi-factor decision-making model [16–18], fuzzy comprehensive evaluation (FCE) [19, 20], AHP [21], BP [22], genetic algorithms (GA) [23], and cellular automatons (CA) [24, 25].

In the IV function, resource condition evaluation is a relatively complex issue. Because the boundary between "suitable" and "unsuitable" is unclear in some circumstances, we decided to use FCE to determine the suitability of land resources. FCE is an integrated evaluation method based on fuzzy mathematics. Unlike the yes-or-no approach of the classical sets, fuzzy sets do not have a definitive, clear boundary for their definitions. They were introduced in the 1960s in an attempt to explain and handle objects with "fuzzy" identities, that is, those

Fig. 3.2 Processes of
integrated fuzzy assessment

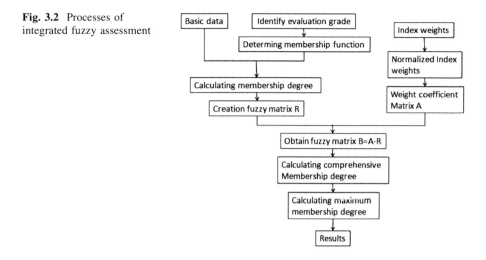

that are not definitive and certain. The application scope of fuzzy mathematics covers almost every aspect and every sector of the national economy [26], particularly, agriculture, forestry, meteorology, environment, geological survey, and the military [27–29].

FCE proves to be an effective solution to generic issues with multiple factors and indicators. With explicit results and a systematic approach, it is capable of solving fuzzy, non-quantitative problems, and it is therefore a good option for land resource suitability assessment [30]. The most-used FCE methods include FCE and fuzzy clustering analysis (FCA). FCE applies primarily to land suitability assessment, land quality assessment, and sustainable development assessment, whereas FCA is used mainly in land type classification and land resource evaluation. Fuzzy mathematics methods have additional applications in land resource evaluation, classification, planning, decision making, forecasting, and control-related studies [26].

The FCE result obtained through calculation was used as a parameter for the calculation of IV function.

See Fig. 3.2 for the FCE operation processes, which include the following steps [31]:

1. Identify the factor domain theory of the evaluation target u.
P is the total number of indicators, as $u = \{u_1, u_2, \ldots, u_p\}$.
2. Identify the domain of evaluation grade v.
$v = \{v_1, v_2, \ldots, v_p\}$ or the grade set. Each grade set corresponds to a fuzzy subset.
3. Creation fuzzy relationship matrix R.
After creating the fuzzy grade subset, we need to quantify each evaluation target against each factor $u_i (i = 1, 2, \ldots, p)$ to determine their respective fuzzy grade subset membership degree, generating the following fuzzy relations matrix:

$$
R = \begin{bmatrix} R| & u_1 \\ R| & u_2 \\ & \cdots \\ R| & u_p \end{bmatrix} = \begin{bmatrix} r_{11} & r_{12} & \cdots & r_{1m} \\ r_{21} & r_{22} & \cdots & r_{2m} \\ \cdots & \cdots & \cdots & \cdots \\ r_{p1} & r_{p2} & \cdots & r_{pm} \end{bmatrix}_{p.m} \tag{3.3}
$$

In matrix R, item r_{ij} in row i and column j indicates the degree of membership of a certain target in fuzzy grade subset v_j evaluated against factor u_i. The performance of the target against factor u_i is described using fuzzy vector $(R|u_i) = (r_{i1}, r_{i2}, \ldots, r_{im})$. In most other evaluation models, it is described with the actual indicator values. In this sense, the FCE requires more information.

Methods for create a membership function, including fuzzy statistics, tripartite, increments, multi-phase fuzzy statistics, a merit-based comparison, an absolute comparison, gathering statistics iterative, priority ordering, a relative comparison, a contrast average, and the function (fuzzy distribution) [32]. Membership degrees can also be determined by directly consulting experts.

4. Determine the weight vector A of the evaluation factors.

We tried to determine the weight vector of the evaluation factors in the FCE: $A = (a_1, a_2, \ldots, a_p)$. Essentially, items a_i in weight vector A are the membership degrees of factor u_i in the fuzzy subset (i.e., the key factors of the evaluation target). The relative priorities of the evaluation indicators can be determined through a number of methods.

5. Obtain the resultant FCE vector.

We used an adequate operator to combine A with the R of the evaluation target, and obtained the resultant FCE vector B of the FCE evaluation items:

$$
A \circ R = (a_1, a_2, \ldots, a_p) \begin{bmatrix} r_{11} & r_{12} & \cdots & r_{1m} \\ r_{21} & r_{22} & \cdots & r_{2m} \\ \cdots & \cdots & \cdots & \cdots \\ r_{p1} & r_{p2} & \cdots & r_{pm} \end{bmatrix} = (b_1, b_2, \ldots, b_m) = B \tag{3.4}
$$

where b_1 is obtained through a calculation conducted between A and column j of R, indicating the overall degree of membership of the evaluation target in fuzzy subset v_j.

6. Analyze the resultant FCE vector.

In practice, the most frequently used method is the maximum membership degree principle. However, in some circumstances, it is not exactly suitable, and causes substantial loss of information or even unreasonable evaluation results. Therefore, we proposed to use the weighted average membership grade method to prioritize the evaluation targets according to their respective grades.

3.3 Technological Advance Intensity Estimation

According to the definition by economists Jacob Schmookler and Edwin Mansfield, technological advance means obtaining more output with the same amount of input; the same amount of output with less input; improving the quality of existing products; or producing new products [33]. From a dialectical perspective, some Chinese scholars have argued that technological advances are a combination of "the development of science" and "the advance of technologies", [34] or the integration of innovation, improvement, and advances of technological elements. Specifically, technological advances include quantitative increases and qualitative improvements to technicians and institutions involved, as well as improvement to tools, equipment, raw materials, techniques, operating processes and methods, skills, organizations, and management practices and means. In addition, it includes quantitative increases and qualitative improvements to researchers and research institutions, research activities, and the latest research results. Technological advance intensity is the extent of technology development in a given period, and it plays an important role in economic development. Technological advance intensity represents latest technology development, and it can be calculated with the MI.

The MI was first proposed in 1953 by Sten Malmquist. In 1982, Caves et al. developed output-oriented and input-oriented MIs [35–37]. However, because Caves et al. did not provide a distance measure function, these MIs are just theoretical concepts and cannot be used in specific empirical studies. Fare et al. broke down the MI model into technical efficiency change and technical progress. Then, they further divided technical efficiency change into scale efficiency changes. Thus, the final MI is expressed as follows:

$$MI = \Delta P \times \Delta S \times \Delta T \tag{3.5}$$

where ΔP is pure technical efficiency change; ΔS is scale efficiency change; and ΔT is technical progress.

Today, the MI is extensively adopted as an important methodology in studies on productivity in the industry, agriculture, finance, medical, public administration and railway sectors (or relevant corporations), or in productivity-related international research projects. Its merit lies in that there is no need to assume a specific production function, which is helpful for the avoidance of function-related errors.

The MI is used in studies regarding efficiency changes of decision-making units in various periods. Its underlying theory is described as follows:

For a technical production possibility set of period t (S^t), technical input is $x \in R_+^N$, technical output is $y \in R_+^M$, and its definition is the following [38]:

$$S^t = \{(x, y) | \text{In the period } t, \quad \text{Input is } x \text{ and output is } y\} \tag{3.6}$$

S^t comprises all the input–output sets feasible in period t. By definition, the distance function is a function created for the purpose of obtaining a meaningful distance based on the production possibility set. We must assume that S^t meets the conditions of some basic axioms.

The output distance function of production activities in period S (x^s, y^s) against the production possibility set of period t (S^t) is defined as follows:

$$D^t(x^s, y^s) = inf\{\theta | (x^s, y^s/\theta) \in S^t\} \tag{3.7}$$

From the definition, it can be inferred that $D^t(x^s, y^s) \leq 1$, $(x^s, y^s) \in S^t$, and $D^t(x^s, y^s) = 1$, meaning that (x^s, y^s) is located at the frontier of the production possibility set S^t. Obviously, in view of the possibility set, production is technically effective.

In essence, the distance function is a description of the ratio between the actual value and the maximum value in period t or the distance between the actual value and the frontier.

In practice, production possibility sets are often generated in accordance with decision-making units (DMUs). Assuming that a total number of K $(k = 1, \ldots)$ DMUs exist, and that each DMU uses N types $(n = 1, \ldots)$ of inputs $x_n^{k,t}$ and obtains M types $(m = 1, \ldots)$ of outputs $y_m^{k,t}$ in period T $(t = 1, \ldots)$.

Based on benchmark technology in periods t and $t + 1$, the MIs of DMU_k are:

$$M_t\left(x^t, y^t, x^{t+1}, y^{t+1}\right) = \frac{D_k^t(x^{t+1}, y^{t+1})}{D_k^t(x^t, y^t)} \tag{3.8}$$

$$M_{t+1}\left(x^t, y^t, x^{t+1}, y^{t+1}\right) = \frac{D_k^{t+1}(x^{t+1}, y^{t+1})}{D_k^{t+1}(x^t, y^t)} \tag{3.9}$$

According to this definition, in order to estimate the productivity of DMU k_0 between t and $t + 1$, it is necessary to obtain the DMUs of 4 different linear planning factors: $D_{k_0}^t(x^t, y^t), D_{k_0}^{t+1}(x^t, y^t)$, $D_{k_0}^t(x^{t+1}, y^{t+1})$, and $D_{k_0}^{t+1}(x^{t+1}, y^{t+1})$ in each $k_0 = 1, 2, \ldots, K$. That is,

$$\left(D_0^t\left(x_{k_0}^t, y_{k_0}^t\right)\right)^{-1} = max \; \theta^{k_0} \tag{3.10}$$

$$\theta^{k_0} y_{k_0 m}^t \leq \sum_{k=1}^{K} z_k^t y_{km}^t$$

$$\sum_{k=1}^{K} z_k^t x_{kn}^t \leq x_{k_0 n}^t$$

$$z_k^t \geq 0$$

$$m = 1, 2, \ldots, M, \quad n = 1, 2, \ldots, N, \quad k = 1, 2, \ldots, K$$

For $D^t_{k_0}(x^{t+1}, y^{t+1})$ can be described as

$$\left(D^t_0\left(x^{t+1}_{k_0}, y^{t+1}_{k_0}\right)\right)^{-1} = max \ \theta^{k_0} \tag{3.11}$$

$$\theta^{k_0} y^{t+1}_{k_0 m} \leq \sum_{k=1}^{K} z^t_k y^t_{km}$$

$$\sum_{k=1}^{K} z^t_k x^t_{kn} \leq x^{t+1}_{k_0 n}$$

$$z^t_k \geq 0$$

$$m = 1, 2, \ldots, M, \quad n = 1, 2, \ldots, N, \quad k = 1, 2, \ldots, K$$

For $D^{t+1}_{k_0}(x^t, y^t)$ is described as

$$\left(D^{t+1}_0\left(x^t_{k_0}, y^t_{k_0}\right)\right)^{-1} = max \ \theta^{k_0} \tag{3.12}$$

$$\theta^{k_0} y^t_{k_0 m} \leq \sum_{k=1}^{K} z^{t+1}_k y^{t+1}_{km}$$

$$\sum_{k=1}^{K} z^{t+1}_k x^{t+1}_{kn} \leq x^t_{k_0 n}$$

$$z^t_k \geq 0$$

$$m = 1, 2, \ldots, M, \quad n = 1, 2, \ldots, N, \quad k = 1, 2, \ldots, K$$

The productivity obtained through this linear planning model is based on the assumption of constant returns to scale (CRS). If we use variable returns to scale (VRS) instead of CRS, the resultant technical productivity could be further broken down into "pure" technical productivity (i.e., technical productivity using VRS production frontier as a benchmark) and efficiency of scale. VRS linear productivity planning is different from CRS productivity planning in that it has one additional restraint:

$$\sum_{k=1}^{K} z^t_k = 1 \tag{3.13}$$

As Because the MIs based on benchmark technologies are symmetric in the economic sense, their geometric mean is defined as an MI:

$$M\left(x^t, y^t, x^{t+1}, y^{t+1}\right) = (M_t \cdot M_{t+1})^{\frac{1}{2}} = \left[\frac{D^t_k(x^{t+1}, y^{t+1})}{D^t_k(x^t, y^t)} \cdot \frac{D^{t+1}_k(x^{t+1}, y^{t+1})}{D^{t+1}_k(x^t, y^t)}\right]^{1/2}$$

$$\tag{3.14}$$

In this book, this MI is used to estimate the intensity of technological advance in the IV function. The obtained result is then used in the calculation of the IV. In practice, adequate DMU should be used to estimate the intensity of technological advances.

3.4 Resource Structure and Distribution Optimization Parameter Calculation

In socio-economic activities, the IV helps increase the value, benefits, and utility of resources by improving the quality or increasing the quantity or concentrations of factors within the same scope of activities. Here, intensive factors include land, labor, capital, technological advances, management, resource allocation, and economies of scale. Typically, land, labor, and capital inputs are called factor inputs, while the combination of all other factors is the TFP. This combination, however, masks changes to individual factors. In 2004, in order to showcase the hidden resource allocation and time-spatial combinations of factors, Zheng Xinqi [9] proposed the concept of "optimized intensive allocation", which was then applied through the combinations of GIS and CA, GA, and data envelopment analysis (DEA) with satisfactory results. Hence, studies on factor structure and allocation of intensive resource use came into focus. Nevertheless, the scope of such studies is limited, and there have been few large-scale case studies. Therefore, resource structures and distribution optimization calculations can be conducted using the following approaches.

3.4.1 Resource Structure Parameter Calculation

There are a number of ways to determine resource exploitation structures. The most intuitive and easy-to-use one is the land Gini coefficient model [8], the best feature of which is the ability to determine differences on a quantitative basis. Based on the Lorenz curve, the land Gini coefficient model is very intuitive. The Gini coefficient for quantitative land use reflects the rationality of quantitative land allocation in specific regions identified for the study. The value range of the land Gini coefficient is 0–1. A smaller value implies increasingly balanced structural ratios, and larger values mean higher degrees of imbalance. Using our previous research [8], we developed a flow chart for the calculation of the land Gini coefficient based on GIS data, as illustrated in Fig. 3.3.

This tool allowed for a quick calculation of the land Gini coefficient, although other tools were available for the same purpose. The result was used in the calculation of the IV function.

Fig. 3.3 Dynamic calculation flow of the basic LGC based on GIS (modified after Zheng) [8]

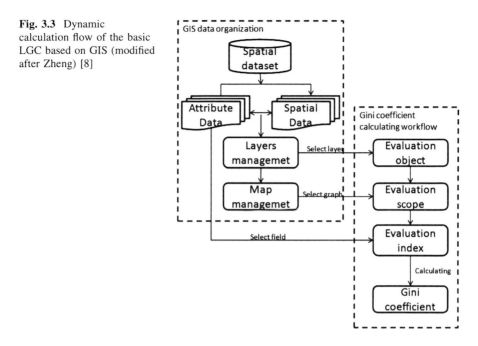

In addition, the CES production function can be used to determine the rationality of the resource structure and allocation. The result obtained can therefore be used in the calculation of the IV function.

3.4.2 Distribution Optimization Parameter Calculation

Distribution optimization parameter calculation has been an aspect often ignored in previous LDR studies. Here, distribution optimization is highlighted because along with the horizontal and vertical evolvement of resource exploitation, every aspect might trigger an unexpected result. Therefore, it should be considered separately in IV studies.

A frequently used indicator in land use spatial distribution or structure studies is the landscape index [39]. Among various landscape index types, some are used to indicate the rationality of spatial distribution [40, 41]. Studies on the rationality of spatial distribution can be conducted at two levels, that is, landscape dimension and patch type, an indicator designed to describe the identities and pattern of spatial distribution of a specific type of patch. In contrast, landscape dimension indicates the overall spatial distribution pattern of the entire landscape while considering the component patches. Although many patch type indicators correspond to those of landscape dimension, they actually have different meanings. Most patch type indicators are used to measure the spatial distribution identities of given patch types, and are therefore known as "fragmental indicators". Alternatively, most landscape dimension indicators are used to measure the overall landscape pattern, and are therefore known as "hybrid indicators". Selecting the right indicators for specific objects in line with their dimensions is critical. However, choosing specific landscape indicators for the distribution optimization calculation should be determined in accordance with the identities of specific research objects. Further research on this aspect has yet to be conducted.

Another distribution optimization parameter can be calculated using the land development density (LDD) [42] index, which consists of building density (BD), architectural dispersion (AD), and construction concentration (CC). For any region, high *BD* and *AD* combined with low *CC* indicates high *LDD* and low development value of the land available. The *BD* factor has the largest effect on the value of *LDD*. Based on this analysis, a mathematical model can be created as follows:

$$LDD = \frac{BD \times (AD)^{1/2}}{CC + 1} \tag{3.15}$$

Before the *LDD* calculation, both dispersion and concentration values should be normalized to eliminate the unit difference. The descending powers treatment was used on the dispersion value to reduce its effect on *LDD*. Adding a constant of 1 to the denominator could reduce the effect of the concentration value on *LDD* and limit the *LDD*'s values to between 0 and 1.

The value obtained was then used in the calculation of the IV function.

3.4.3 Time Change Parameter Calculation

Time and location are two dimensions that describe the order of things. The IV can reflect intensive changes in terms of time and location. Built on the basis of LDR, IV implies time accumulation. All human activities have two aspects: accumulation and dynamic changes [43].

Chronological data are values of one or more variables in a given period, whereas cross-sectional data are those of several units or objects collected at the same point in time. Panel data are data of the same section collected several times during the course of study [44] with two dimensions: chronological order and sections. Panel data proves to be more valuable through the combination of the chronological order and section attributes, because the variables are now more diversified and less collinear.

Panel data allow the reflection of evaluation targets in a chronological order. A chronological order ratio is obtained through calculation, and the result may be used in the calculation of the IV function.

In summary, IV is meaningful in time and spatial dimensions, and it has structural meanings. Based on this parameter calculation, it is possible to consider a specific value in line with each situation and apply it to a summarization calculation similar to those of factor inputs and TFP or to calculate the parameters separately and use the results in the calculation of the IV function.

References

1. Yang, Shengbiao. 2009. On the applicability of CD function, the CES function and VES function. *Statistics and Decision* 5: 153–155.
2. Lv, Zhendong, Jue Guo, and Youmin Xi. 2009. Econometric estimate and selection on China energy CES production function. *China Population Resources and Environment* 19(4): 156–160.
3. Berndt, E.R. 1976. Reconciling alternative estimates of the elasticity of substitution. *The Review of Economics and Statistics* 58(1): 59–68.
4. Xu, Zhuoshun. 2009. *Computable general equilibrium (CGE) model: the modeling principle, parameter estimation methods and application.* Ph.D. thesis, Jilin University.
5. Yuan, Qinghe, Wenfu Fu, and Huijun Sun. 2001. A study on assessment index system of regional resource conditions. *Journal of Shandong University of Science and Technology (Natural Science)* 20(3): 91–93.
6. Yu, Xiaofen, and Dai Fu. 2004. Review of the multi index comprehensive evaluation method. *Statistics and Decision* 11(179): 119–121.
7. Wang, Jiating, and Liang Zhao. 2009. An empirical study on provincial urbanization efficiency. *Journal of Tongji University (Social Science Section)* 4: 44–50.
8. Zheng, Xinqi, Tian Xia, Xin Yang, et al. 2013. The land Gini coefficient and its application for land use structure analysis in China. *PLoS ONE* 8(10): e76165. doi:10.1371/journal.pone.0076165.
9. Zheng, Xinqi. 2004. *Optimize the allocation of intensive use evaluation of urban land-theory, methods, techniques, empirical.* Beijing: Science Press.
10. Hopkins, L. 1997. Methods for Generating land suitability maps: a comparative evaluation. *Journal for American Institute of Planners* 34(1): 19–29.
11. Collins, M.G., F.R. Steiner, and M.J. Rushman. 2001. Land-use suitability analysis in the United States: historical development and promising technological achievements. *Environmental Management* 28(5): 611–621.
12. Deng, Qingchun. 2008. *GIS-based agricultural land suitability evaluation: a case study of longquanyi city.* Master's Thesis, Sichuan University.
13. Ni, Shaoxiang. 1999. *Introduction to land type and land evaluation(Second Edition).* Beijing: Higher Education Press.
14. Pereira, J.M.C., and L. Duckstein. 1993. A multiple criteria decision-making approach to GIS-based land suitability evaluation. *International Journal of Geographical Information Systems* 7(5): 407–424.
15. Malczewski, J. 1996. AGIS-based approach to multiple criteria group decision-making. *International Journal of Geographical Information Systems* 10(8): 955–971.
16. Carver, S.J. 1991. Integrating multi-criteria evaluation with geographical information systems. *International Journal of Geographical Information Systems* 5(3): 321–339,
17. Banai, R. 1993. Fuzziness in geographic information systems: contributions from the analytic hierarchy process. *International Journal of Geographical Information Systems* 7(4): 315–329.
18. Malczewski, J. 1999. *GIS and multi-criteria decision analysis.* New York: Wiley.
19. Zadeh, L.A. 1965. Fuzzy Sets. *Information and Control* 8(3): 338–353.
20. Qiu, Bingwen, Tianhe Chi, Qinmin Wang, et al. 2004. Application of GIS and its prospect in land suitability assessment. *Geography and Geo-Information Science* 20(5): 20–23, 44.

21. Zhang, Qiulin, Baolian Li, Dongmin Li, et al. 2009. Evaluation on suitability of needed reclaiming land in Xinhe mining area by AHP. *Guizhou Agricultural Sciences* 37(5): 102–104.
22. Zhou, J., and D.L. Civco. 1996. Using genetic learning neural networks for spatial decision making in GIS. *Photogrammetric Engineering and Remote Sensing* 62(11): 1287–1295.
23. Krzanowski, R., J. Raper, and M.D. Berg. 2001. *Spatial evolutionary modeling*. UK: Oxford University Press.
24. Wu, F. 1998. Simland: a prototype to simulate land conversion through the integrated GIS and CA with AHP derived transition rules. *International Journal of Geographical Information Science* 12(1): 63–82.
25. Opensha, W.S., and R.J. Abrahart. 2000. *Geo computation*, 187–217. London: Taylor &Francis.
26. Li, Xican, Jing Wang, and Xiaomei Shao. 2009. Progress in the application of fuzzy mathematics methods in China land resource evaluation. *Progress in Geography* 28(3): 409–416.
27. Liu, Jinda, and Mengda Wu. 2000. *Fuzzy theory and Its Applications (Second Edition)*. National Defense Science and Technology University Press.
28. Wang, Yaqiu, and Baoyuan Zheng. 1996. Fuzzy comprehensive evaluation of ecological suitability degree. *Fujian Environment* 15(5): 7–10.
29. Changzhong Fan. 1995. *Urban ecological quality evaluation in Guangdong province*. Master's Thesis, Sun Yat-Sen University.
30. Yu, Sheng. 2007. *The application of fuzzy judgment method in the regional industrial land-use plans suitability analysis*. Master's Thesis, Chinese Academy of Agricultural Sciences.
31. Li, Shiyong. 2004. *Engineering fuzzy mathematics and its application*. Harbin: The Polytechnic University of Harbin Press.
32. Su, Weihua. 2000. *Research on Theory and Method of Multi Index Comprehensive Evaluation*. Ph.D. thesis, Xiamen: Xiamen University.
33. Peng, Hong, and Sharen Zhang. 2011. Economic interpretation of the progress of science and technology. *Science Technology and Industry* 11(11): 143–150.
34. Cai, Yongsheng. 1994. Discussion on the basic concept of the progress of science and technology. *Academic Journal Graduate School Chinese Academy of Social Sciences* 2: 66–72.
35. Orea, Luis. 2002. Parametric decomposition of a generalized Malmquist productivity index. *Journal of Productivity Analysis* 18: 5–22.
36. Xu Zhao. 2007. Productivity evaluation and empirical analysis of China's insurance industry based on Malmquist index. *Industrial Economics Research* 2: 8–13.
37. Caves, D.W., L.R. Christensen, and W.E. Diewert. 1982. Multilateral comparisons of output, input, and productivity using superlative index numbers. *Economic Journal* 92(365): 73–86.
38. Zhang, Xiangsun, and Binwei Gui. 2008. The analysis of total factor productivity in China: a review and application of Malmquist index approach. *The Journal of Quantitative & Technical Economics* 6: 111–122.
39. Zheng, Xinqi, and Meichen Fu. 2010. *Landscape pattern spatial analysis technology and its application*. Beijing: Science Press.
40. Zheng, Yu., Yecui Hu, Yansui Liu, et al. 2005. Spatial analysis and optimal allocation of land resources based on land suitability evaluation in shandong province. *Transactions of the CSAE* 21(2): 60–65.
41. Luo, Ding, Xu Yueqing, Xiaomei Shao, et al. 2009. Advances and prospects of spatial optimal allocation of land use. *Progress in Geography* 28(5): 791–797.
42. Li, Lihua. 2008. Study on the land economical and intensive utilization of rural-urban fringe in beijing, China Master's Thesis. China University of Geosciences in Beijing.
43. Zheng, Xinqi, Hu, Yecui, et al. 2013. An evaluation approach for regional construction land intensive use based on binding static & dynamic methods. *Urbanization and Land Use* 1: 3–8.
44. Li, Zinai, and Wenqing Pan. 2000. *Econometrics (Third Edition)*. Beijing: Higher Education Press.

Chapter 4
Computing Tools

Abstract Computing tools are necessary for IV models. Both Excel and GIS software have been developed to be useful tools for complicated calculation in IV model. The Excel-based computing tool enable a semi-automated computation with the help of VBA, while the GIS-based one can realize entire computing and fitting processes of intensive resource use in a particular region and in time-spatial dimensions. The details of tools design, especially of the GIS-based computing tool, are explained in this chapter. A range of features, such as data sorting and management, resource condition suitability assessment, IV curve generation, etc., are provided in the computing tools.

Keywords Computing tools · GIS · Excel · Tool design · Tool features

IV functions involve complex calculations. To ensure higher efficiency, it is necessary to have computing tools. In the course of theoretical research, we attempted to develop two computing tools, one based on excel and the other based on GIS.

4.1 Excel-Based IV Computing Tool

Technological advances are calculated using Microsoft Excel [1], whereas IV parameters are analyzed with the multiple regression model. Key parameters involved in IV functions include resource conditions, labor input, capital input, technological advances, and various elasticity coefficients and parameters, among which, resource conditions (R), technological advances (T), resource structure and distribution optimizations (M), elasticity coefficients (α and β), and errors (δ) must be calculated. Through resource condition evaluation and the follow-on fuzzy judgment, a specific value is obtained for further calculation. Technological advance is calculated using the MI, and the result is used in the follow-on

X. Zheng et al., *Intensive Variable and Its Application*, SpringerBriefs in Geography, DOI: 10.1007/978-3-642-54873-4_4, © The Author(s) 2014

calculation. Resource conditions, labor input, and capital input can be obtained depending on relevant attributes of the evaluation target and used in the follow-on calculation. Other items, that is, δ, α, and β, must be calculated through the multiple regression model.

The Excel-based computing tool provides a data sheet, a computation sheet, and a result sheet to enable a semi-automated computation with the help of VBA. Because the Excel-based computing tool was developed primarily to assist in the study, it is not very convenient despite a generally satisfactory result. To save space, we do not further detail this particular aspect.

4.2 GIS-Based IV Computing Tool

4.2.1 Introduction

Excel is not sufficient to solve all calculation problems and time-spatial problems of the IV, and thus, a new computing tool is necessary. GIS is a technical system developed with the support of specific computer hardware and software for the collection, storage, management, computing, analysis, display, and description of geographic distribution data. GIS processes and manages data regarding specific objects in different geographic locations and the relations between these objects (including spatial positioning data, graphic data, and membership data) to analyze and process phenomena and procedures in a particular region and to solve complicated planning, decision-making, and management problems. Developed by the Environment System Research Institute (ESRI), ArcGIS is one of the most extensively used GIS-based applications in the world. With a powerful spatial information processing capability, ArcGIS can be used for IV calculation and time-spatial evaluation. In this section, we introduce a number of ArcGIS 9.3 platform-based computing methods for the calculation of IV.

With this integrated computing tool, it is possible to describe IV theory to enable the entire computing and fitting processes of intensive resource use in a particular region. The combination of IV theory with GIS theory allows the processing of IV in time-spatial dimensions. This tool enables computer-aided calculation of IV of an entire region or any of its subordinate jurisdictions or higher visibility portions.

4.2.2 Design

The IV computing tool is developed on the basis of GIS technologies to provide a range of capabilities, including resource suitability assessment, technological advance calculation, IV fitting, and resource intensification analysis.

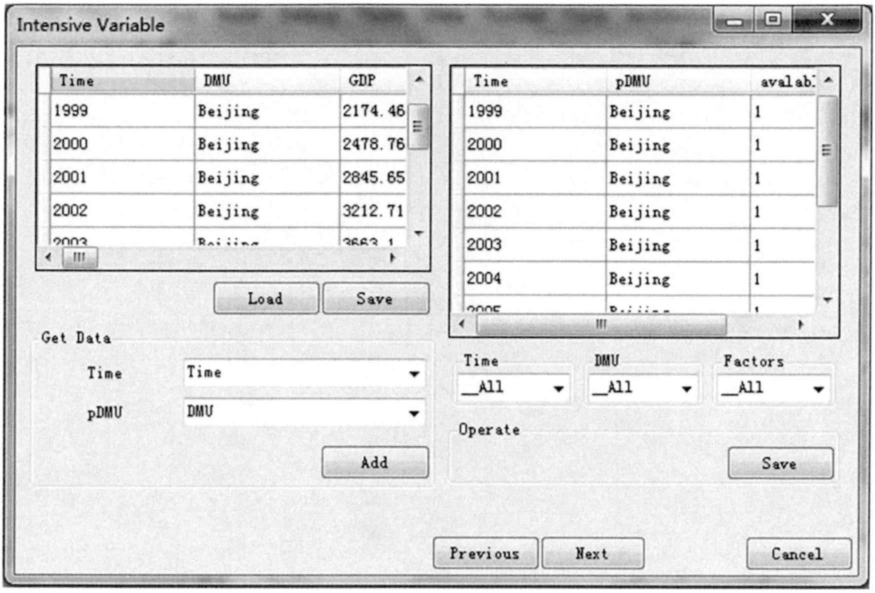

Fig. 4.1 Data retrieval dialogue box

Furthermore, the tool is developed within the Microsoft.NET framework, using Microsoft Visual Studio 2010 as the basic development environment and Microsoft Visual C# as the programming language. In addition, it leverages ArcGIS's secondary development platform ArcEngine, primarily its Shape files for data aggregation.

The hardware environment of the application includes a currently popular configuration of a 250 GB+ idle hard drive, a quad-core processor, 2 GB memory, and Windows 7.

4.2.3 Features

IV tools provide a range of features, including data sorting and management, geographic location data retrieval, resource condition suitability assessment, technological advance calculations, IV model fitting, capital and labor input elasticity coefficient calculation, IV curve generation, and presentation of evaluation results in the diagram.

1. Data sorting and management

Data management is conducted on the GIS secondary development platform. Particularly, the data entry feature allows the opening and browsing of data intended for IV fitting for a specific region as well as the import of data into the tool for follow-on calculations. Figure 4.1 shows the data entry dialogue box.

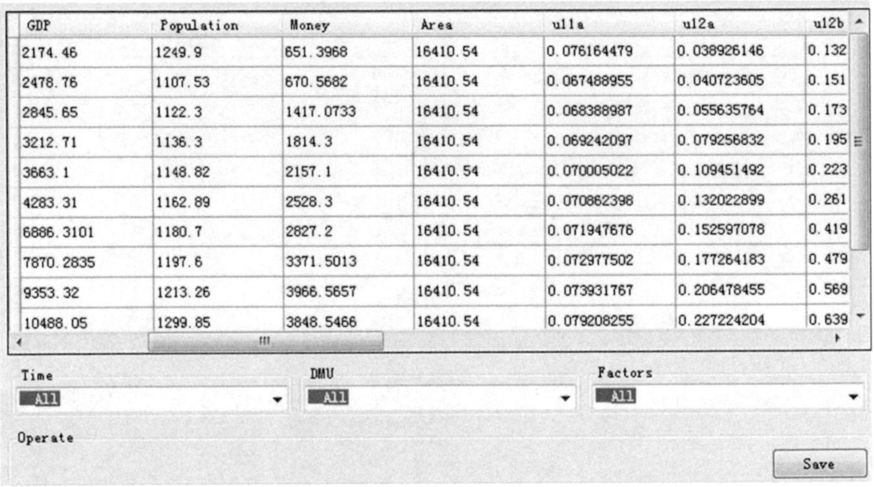

GDP	Population	Money	Area	ul1a	ul2a	ul2b
2174.46	1249.9	651.3968	16410.54	0.076164479	0.038926146	0.132
2478.76	1107.53	670.5682	16410.54	0.067488955	0.040723605	0.151
2845.65	1122.3	1417.0733	16410.54	0.068388987	0.055635764	0.173
3212.71	1136.3	1814.3	16410.54	0.069242097	0.079256832	0.195
3663.1	1148.82	2157.1	16410.54	0.070005022	0.109451492	0.223
4283.31	1162.89	2528.3	16410.54	0.070862398	0.132022899	0.261
6886.3101	1180.7	2827.2	16410.54	0.071947676	0.152597078	0.419
7870.2835	1197.6	3371.5013	16410.54	0.072977502	0.177264183	0.479
9353.32	1213.26	3966.5657	16410.54	0.073931767	0.206478455	0.569
10488.05	1299.85	3848.5466	16410.54	0.079208255	0.227224204	0.639

Time **All**   DMU **All**   Factors **All**

Operate   Save

Fig. 4.2 Dialogue box after data import (in the case of Beijing)

Click 【Add】 to add data to the calculation data set. Adjust evaluation units, time, and indicators to view different data subsets. See Fig. 4.2 for the view after data import.

2. Resource Condition Suitability Assessment

There are many methods of conducting a resource condition suitability assessment. This tool uses a fuzzy model to assess the suitability of relevant resource conditions, such as land. After the import of data, the resource condition suitability assessment phase starts (Fig. 4.3).

In the assessment indicator system, identify names, criteria, types, weights of indicators, and names and weights of secondary indicators. Click 【Calculate】 to start the calculation. See Fig. 4.4 for the results of the resource condition suitability assessment.

3. Technological Advance Calculation

After the completion of resource assessment, conduct a technological advance calculation with Deap V2.1 [2]. See Fig. 4.5 for the calculation interface.

In the dialogue box, first select whether each factor is an input or output, shown to the right of each factor. Specifically, "NONE" means not applicable; "IN" means input; and "OUT" means output. Click 【Calculate】 to calculate the technological advancement of a given year. See Fig. 4.6 for the results.

4. IV Model Fitting

After the completion of the technological advancement calculation, continue to the IV model fitting dialogue box (Fig. 4.7).

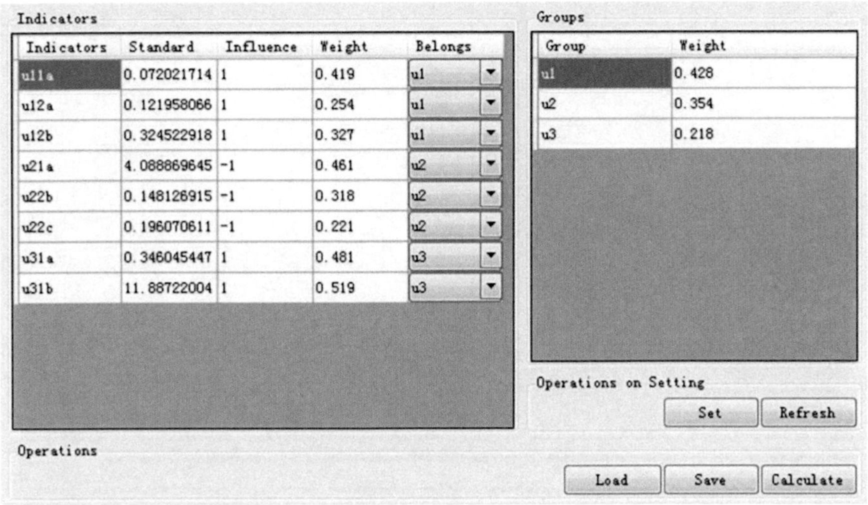

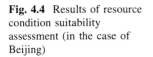

Indicators	Standard	Influence	Weight	Belongs		Group	Weight
u11a	0.072021714	1	0.419	u1 ▼		u1	0.428
u12a	0.121958066	1	0.254	u1 ▼		u2	0.354
u12b	0.324522918	1	0.327	u1 ▼		u3	0.218
u21a	4.088869645	-1	0.461	u2 ▼			
u22b	0.148126915	-1	0.318	u2 ▼			
u22c	0.196070611	-1	0.221	u2 ▼			
u31a	0.346045447	1	0.481	u3 ▼			
u31b	11.88722004	1	0.519	u3 ▼			

Operations on Setting

[Set]   [Refresh]

Operations

[Load]   [Save]   [Calculate]

Fig. 4.3 Resource condition suitability assessment

Fig. 4.4 Results of resource condition suitability assessment (in the case of Beijing)

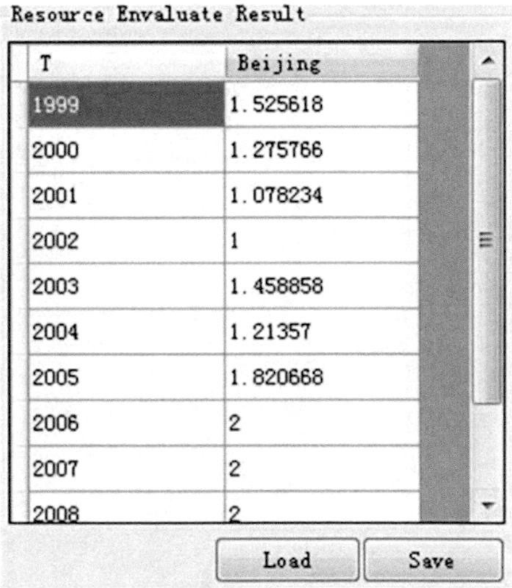

Resource Envaluate Result

T	Beijing
1999	1.525618
2000	1.275766
2001	1.078234
2002	1
2003	1.458858
2004	1.21357
2005	1.820668
2006	2
2007	2
2008	2

[Load]   [Save]

The values of R and T in the IV function have been obtained through previous calculations, whereas those of Q, S, K, and L must be selected from the data available. Click 【Calculate】 to calculate elasticity coefficients α and β of K (capital input) and L (labor input). See Fig. 4.8 for the results.

Use the values of α and β obtained in the IV function to calculate the IV model of the given year. When $\alpha \geq \frac{\alpha+\beta}{2}$, the evaluation target is under intensive development; however, when $\beta \geq \frac{\alpha+\beta}{2}$, the evaluation target is under extensive development.

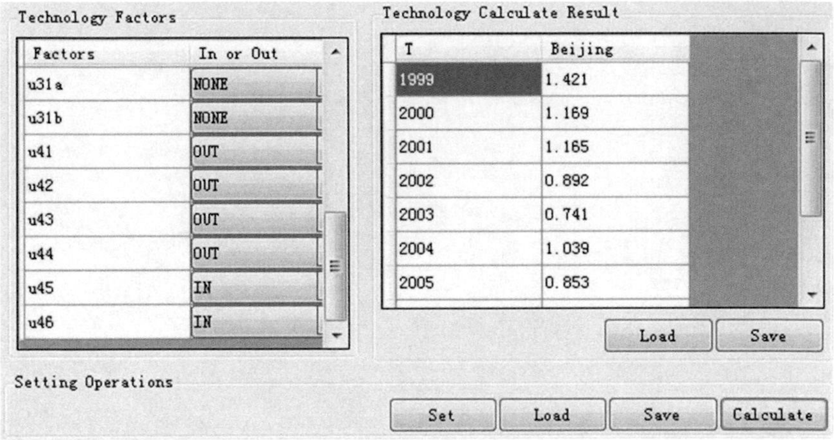

Technology Factors

Factors	In or Out
u31a	NONE
u31b	NONE
u41	OUT
u42	OUT
u43	OUT
u44	OUT
u45	IN
u46	IN

Technology Calculate Result

T	Beijing
1999	1.421
2000	1.169
2001	1.165
2002	0.892
2003	0.741
2004	1.039
2005	0.853

Load   Save

Setting Operations

Set   Load   Save   Calculate

Fig. 4.5 Technological advance calculation dialogue box

Fig. 4.6 Results of technological advance calculation (for Beijing, 1999–2007)

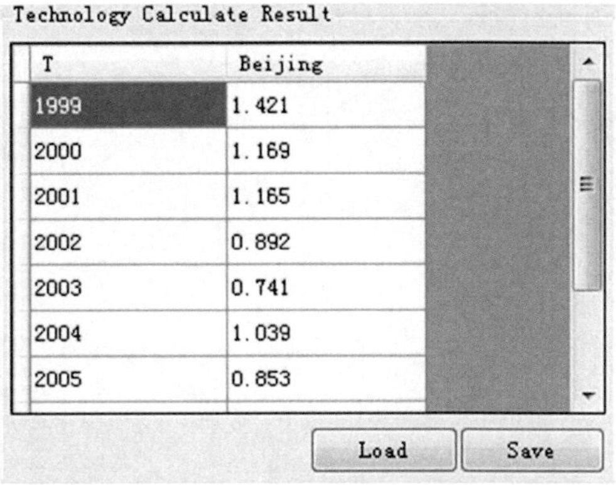

Technology Calculate Result

T	Beijing
1999	1.421
2000	1.169
2001	1.165
2002	0.892
2003	0.741
2004	1.039
2005	0.853

Load   Save

Fig. 4.7 IV model fitting dialogue box

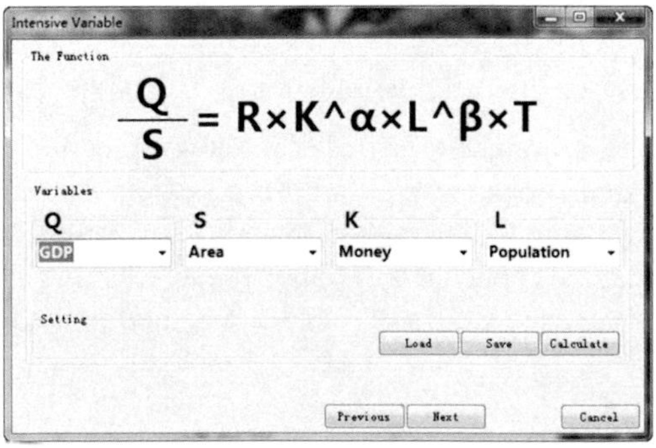

Intensive Variable

The Function

$$\frac{Q}{S} = R \times K^{\wedge}\alpha \times L^{\wedge}\beta \times T$$

Variables

Q	S	K	L
GDP	Area	Money	Population

Setting

Load   Save   Calculate

Previous   Next   Cancel

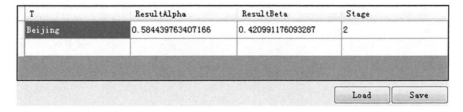

T	ResultAlpha	ResultBeta	Stage
Beijing	0.584439763407166	0.420991176093287	2

Load   Save

Fig. 4.8 IV model fitting results

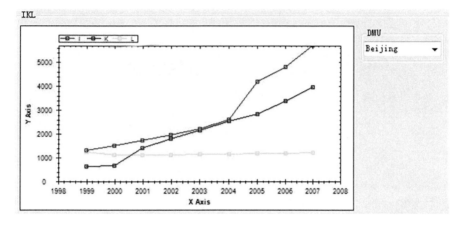

Fig. 4.9 IV curve mapping dialogue box

5. IV Curve Mapping

In addition to the capital input (K) curve, labor input (L) curve, and IV (I) curve, the feature also enables the mapping of curves modified by resource conditions and technology, that is, curves that reflect total product (I_L) increasing only labor input (all other factors constant), total product (I_K) increasing only capital input (all other factors constant), MP (I'_K) and AP (A_K) per unit of capital input, and MP (I'_L) and AP (A_L) per unit of labor input.

Next, advance to the IV curve mapping dialogue box (Fig. 4.9). The DMU drop-down list allows the user to switch the items to display.

6. Resource Intensification Analysis

Intensive development of the evaluation target can be grouped into low-efficiency, medium-efficiency, intensive, over-intensive, and post-intensive use stages.

The result of intensive development obtained through calculation can be revised through curve analysis. The results can also be displayed on the map. See Fig. 4.10.

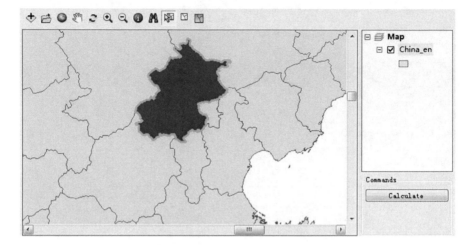

Fig. 4.10   Results of intensive development

7. Time-Spatial Analysis of Resource Intensification

In the GIS dialogue box, select an evaluation object in the DMU drop-down list or directly in the graphic interface. You can either select a single object for overall evaluation or the subordinate factors of a region for calculation in chronological order while reflecting the spatial changes. In other words, the tool enables evaluation of all land resources of a particular region or that of an integral land plot.

References

1. Yin, CongChun. 2012. The application of excel programming in economic management. *China Management Information* 15(4): 36–37.
2. The University of Queensland. 2013. A data envelopment analysis (computer) program. http://www.uq.edu.au/economics/cepa/deap.php, 12 Dec 2013.

Chapter 5
Intensive Evaluation of Regional Land Use

Abstract On the basis of IV theory and computing tools discussed in previous chapters, this chapter studies intensive land use evaluation on region scale. A case study of Henan Province in China during 1999–2008 is operated for IV model fitting for the subordinate cities in line with the general land Gini coefficient of the city. After data collection and pre-processing, 18 subordinate areas of Henan were decided to be the evaluation units, and indicators including GDP, land area, resource condition suitability assessment, total fixed asset investment, population, and technological advance are selected for a simplified model. Results show that most land in Henan province is currently in the transition from the medium-intensity stage to the intensive stage and has yet to progress into the fully intensive stage. The case study proves the capability of IV model in reflecting the economic development status of a particular region. The IV theory is both rooted in and builds on the LDR theory, which is an increasing step function of the former.

Keywords Case study · Intensive land use evaluation · Region scale · Henan

Regional land use involves all the land available in a particular region. IV for regional land use are variables that enable the increase of value, benefits, and utility of land as a result of human activities within a particular region. This chapter studies intensive land use using a case study of Henan Province in China for IV model fitting for the subordinate cities in line with the general land Gini coefficient of the city. Specifically, resource quantity is the total land area within the jurisdiction of the province, and the scope of resource condition suitability assessment is the entire province.

5.1 Data and Pre-processing

Most of the data used in this study were obtained from the national database[1] on the official website of the National Bureau of Statistics, with the remaining data abstracted from the statistical yearbooks of the provinces involved. Initial data

---

[1] http://data.stats.gov.cn/index

X. Zheng et al., *Intensive Variable and Its Application*, SpringerBriefs in Geography, 43
DOI: 10.1007/978-3-642-54873-4_5, © The Author(s) 2014

processing was first conducted using Excel. Then, leveraging the attribute coupling function of GIS, these data were combined with spatial data into specific datasets available for use with computing tools. Spatial data include basic geographic information elements and IV data positioning foundations. The 1980 Xi'an coordinate system (CS) was adopted as the plane CS along with a Gauss-Krüger projection and the 6° zone division. Spatial data were obtained from the 1:400 shared data[2] of the National Geomatics Center of China (NGCC). The FCE data and MI calculation were beyond the scope of the dataset setup.

5.2 Evaluation Units and Indicator Selection

Henan Province was selected as the target of intensive land use study, and its 18 subordinate prefectural cities were selected as the evaluation units.

In theory, the indicators of the IV model were identified in line with the specific aspects of the research target. In our case of regional land use evaluation, we selected a simplified model to streamline the calculation process, which allowed us to combine technological advances with distribution optimization into a single variable (T):

$$I = R \times K^{\alpha} \times L^{\beta} \times T \tag{5.1}$$

See Table 5.1 for the socioeconomic, population, resource, and technology indicators selected. Because a resource condition assessment requires year pair data (i.e., 2 years), data from 1997–2008 were collected to simulate IV models of 1999–2008.

One of the resource condition suitability assessment-related indicators is land area available for urban development. In the technological advancement category, technological input refers to staffing and R&D expenses of large and medium enterprises in Henan Province, and technological output refers to new product output and Chinese patents granted to industrial enterprises above the designed scale. The sources of these socioeconomic, land area, and population data include the statistics authorities of the central and relevant provincial statistics authorities, whereas technological advance data are quoted from past years' China Statistical Yearbook of Science and Technology.[3]

Based on the result of initial data processing, a simulation was made on intensive land use of 18 prefectural cities in Henan Province during 1999–2008.

---

[2] http://ngcc.sbsm.gov.cn/article/sjcg/dtxz/e/
[3] http://tongji.cnki.net/kns55/navi/HomePage.aspx?id=N2013010081&name=YBVCX&floor=1

Table 5.1 IV indicators for regional land use

Indicator	Q	S	R	K	L	T
Meaning	GDP (RMB 10 k)	Land area (km²)	Resource condition suitability assessment with FCE	Total fixed asset investment (RMB 100 mn)	Population (10 k)	Technological advance

5.3 Indicator Calculation

5.3.1 Resource Condition Suitability Assessment

Resource condition indicator values were obtained using a suitability assessment with the multi-layer FCE model previously described.

1. Identify factor sets

Assuming that advanced factors include utilization intensity (u_1), land consumption in economic growth (u_2), and the land elasticity index (u_3), primary factors include population density (u_{11}), fixed asset investment per land area unit (u_{12}), GDP per land area unit (u_{13}), incremental construction land area consumed by population growth (u_{21}), incremental construction land area consumed by regional GDP growth (u_{22}), the elasticity coefficient of population and construction land area growth (u_{31}), and the elasticity coefficient of regional GDP and construction land area growth (u_{32}). Hence, the factor indicator set is

$$U = \left\{ \begin{array}{l} u_{11}, u_{12}, u_{13} \\ u_{21}, u_{22}, u_{23} \\ u_{31}, u_{32} \end{array} \right\} \tag{5.2}$$

2. Setup the comment set

Assume that v_1 indicates that the result is "suitable" and v_2 indicates that it is "unsuitable", the comment set is

$$V = \{v_1, v_2\} \tag{5.3}$$

3. Develop the membership function

Compare the indicator values of individual cities with the average values of Henan Province over the years. For positive factors, if the indicator value of a city is larger than the provincial average, the result is suitable; otherwise, it is unsuitable. For negative factors, if the indicator value of a city is larger than the provincial average, the result is unsuitable; otherwise, it is suitable. The membership function is described as follows:

The membership function of positive factor indicators is

$$r_{ijv_1} = \begin{cases} 1 & u_{ij} \geq u_0 \\ 0 & u_{ij} < u_0 \end{cases} \tag{5.4}$$

$$r_{ijv_2} = 1 - r_{v_1}$$

The membership function of negative factor indicators is

$$r_{ijv_1} = \begin{cases} 0 & u_{ij} \geq u_0 \\ 1 & u_{ij} < u_0 \end{cases} \tag{5.5}$$

$$r_{ijv_2} = 1 - r_{v_1}$$

where r_{ijv_1} and r_{ijv_2} are memberships of primary factor item j of advanced factor item i in the comment set; u_{ij} is the primary factor item j of advanced factor item i; $i = 1, 2, \ldots, n$, $j = 1, 2, \ldots, n$; and u_0 is the average value of Henan Province for each year.

The result is a total number of i primary membership matrices:

$$R_i = \begin{pmatrix} r_{11} & r_{12} & \cdots & r_{1j} \\ r_{21} & r_{22} & \cdots & r_{2j} \\ \vdots & \vdots & \vdots & \vdots \\ r_{i1} & r_{i2} & \cdots & r_{ij} \end{pmatrix} \tag{5.6}$$

Using Zhengzhou as an example, its primary membership matrix results are shown in Table 5.2.

4. Identify the weights of factor sets

Determine the factor weights using the Delphi method [1]. The weights of the primary factors are:

$$A_1 = (0.419, 0.254, 0.327)$$
$$A_2 = (0.461, 0.318, 0.221)$$
$$A_3 = (0.481, 0.519)$$

The weights of the advanced factors are

$$A = (0.428, 0.354, 0.218)$$

Table 5.2 Resource condition FCE primary membership calculation results (Zhengzhou)

Zhengzhou, Henan province

Membership	1999	2000	2001	2002	2003	2004	2005	2006	2007	2008
u11	1	1	1	1	1	1	1	1	1	1
u12	1	1	1	1	1	1	1	1	1	1
u13	1	1	1	1	1	1	1	1	1	1
u21	0	1	0	0	0	0	1	0	0	0
u22	0	0	0	0	0	0	0	0	0	0
u23	1	1	0	0	0	1	0	0	1	0
u31	0	0	1	0	0	0	0	1	1	0
u32	0	0	1	1	0	0	0	1	1	0

5. Conduct a fuzzy compound matrix calculation

First, conduct a primary judgment.

$$B_i = A_i \circ R_i \quad (i = 1, 2, 3) \tag{5.7}$$

$$b_{ij} = min\left[1, \sum_{i=1}^{n} (a_{ij} \cdot r_{ij})\right] \quad (j = 1, 2, 3 \text{ or } 1, 2) \tag{5.8}$$

where B_i is the single-factor judgment of U_i; a_{ij} is the weight of primary factor j of advanced factor i.

See Table 5.3 for the results of the primary judgment.

Then, conduct a secondary judgment. The general judgment matrix R for advanced factor set U is

$$\mathbf{R} = \begin{pmatrix} \mathbf{B_1} \\ \mathbf{B_2} \\ \vdots \\ \mathbf{B_k} \end{pmatrix} = \begin{pmatrix} \mathbf{A_1 \circ R_1} \\ \mathbf{A_2 \circ R_2} \\ \vdots \\ \mathbf{A_k \circ R_k} \end{pmatrix} \tag{5.9}$$

The result of the secondary comprehensive judgment is $B = A \circ R$. Set B as the result of the comprehensive judgment of the factor set, as shown in Table 5.4.

6. Determine the final resource condition values

When resource conditions are kept constant, their value in the IV model is 1. If the resource conditions are variable, their value in the IV model is set between 1 and 2 to reflect the features of resource condition changes and to differentiate the case from that of constant resource conditions. Based on the result of this FCE, the value of the resource conditions in the variable case is obtained by adding 1 to the suitable membership of the comment set. See Table 5.5 for specific results.

5.3.2 Technological Advances

If relevant data are available, applicable technological inputs and outputs are selected for the calculation of technological advances with the MIs.

Because technology data of all cities in Henan are available, technological advances of 1999–2008 were calculated using the MIs (see Table 5.6).

5.3.3 Other Indicators used in the Model

In order to simplify the calculations, we ignored the errors here.

In the intensive land use model of the region, the IV is GDP per land area unit, or RMB 10 k/km^2. Other indicators are used directly for the IV model fitting.

Table 5.3 FCE primary judgment results (Zhengzhou)

Zhengzhou, Henan province

Membership	1999			2000			2001			2002			2003			2004			2005			2006			2007			2008		
U1	1.00	0.00	0.00	1.00	0.00	0.00	1.00	0.00	0.00	1.00	0.00	0.00	1.00	0.00	0.00	1.00	0.00	0.00	1.00	0.00	0.00	1.00	0.00	0.00	1.00	0.00	0.00	1.00	0.00	0.00
U2	0.22	0.78	0.00	1.00	0.00	0.00	0.46	0.54	0.00	1.00	0.00	0.00	1.00	0.00	0.00	1.00	0.00	0.00	1.00	0.00	0.00	0.54	0.46	0.00	1.00	0.00	0.00	1.00	0.00	0.00
U3	0.00	1.00	0.00	0.00	1.00	0.00	0.00	1.00	0.00	0.00	1.00	0.00	0.00	0.00	1.00	0.00	0.00	1.00	0.00	0.00	1.00	0.00	0.00	1.00	0.00	0.00	1.00	0.00	0.00	1.00

Table 5.4 FCE membership calculation results

Membership	1999		2000		2001		2002		2003		2004		2005		2006		2007		2008	
Zhengzhou	0.506	0.494	0.782	0.218	0.591	0.409	0.428	0.572	0.646	0.354	0.646	0.354	0.428	0.572	0.619	0.381	0.428	0.572	0.782	0.218
Kaifeng	0.537	0.463	0.595	0.405	0.895	0.105	0.891	0.109	0.623	0.377	0.456	0.544	0.638	0.362	0.891	0.109	0.673	0.327	0.319	0.681
Luoyang	0.467	0.533	0.000	1.000	0.354	0.646	0.000	1.000	0.218	0.782	0.572	0.428	0.572	0.428	0.444	0.556	0.140	0.860	0.603	0.397
Pingdingshan	0.494	0.506	0.751	0.249	0.673	0.327	0.891	0.109	0.751	0.249	0.673	0.327	0.751	0.249	0.891	0.109	0.891	0.109	0.891	0.109
Anyang	0.506	0.494	0.398	0.602	0.595	0.405	0.891	0.109	0.616	0.384	0.751	0.249	0.179	0.821	0.673	0.327	1.000	0.000	1.000	0.000
Hebi	0.887	0.113	0.666	0.334	0.587	0.413	0.778	0.222	0.751	0.249	0.751	0.249	0.673	0.327	1.000	0.000	1.000	0.000	0.428	0.572
Xinxiang	0.778	0.222	0.398	0.602	0.258	0.742	0.319	0.681	0.179	0.821	0.397	0.603	0.179	0.821	0.669	0.331	0.428	0.572	0.782	0.218
Jiaozuo	0.506	0.494	0.619	0.381	0.506	0.494	0.887	0.113	0.428	0.572	0.646	0.354	0.782	0.218	0.895	0.105	0.541	0.459	0.428	0.572
Puyang	1.000	0.000	1.000	0.000	1.000	0.000	1.000	0.000	0.646	0.354	1.000	0.000	1.000	0.000	1.000	0.000	1.000	0.000	0.782	0.218
Xuchang	1.000	0.000	0.428	0.572	0.782	0.218	0.922	0.078	0.428	0.572	1.000	0.000	1.000	0.000	1.000	0.000	0.837	0.163	0.782	0.218
Luohe	0.619	0.381	0.782	0.218	0.782	0.218	0.817	0.183	1.000	0.000	1.000	0.000	0.533	0.467	0.541	0.459	0.782	0.218	1.000	0.000
Sanmenxia	0.113	0.887	0.000	1.000	0.241	0.759	0.467	0.533	0.467	0.533	0.572	0.428	0.572	0.428	0.304	0.696	0.304	0.696	0.304	0.696
Nanyang	0.572	0.428	0.572	0.428	0.000	1.000	0.572	0.428	0.572	0.428	0.572	0.428	0.782	0.218	0.218	0.782	0.241	0.759	0.354	0.646
Shangqiu	0.534	0.466	0.751	0.249	0.533	0.467	0.456	0.544	0.370	0.630	0.860	0.140	0.673	0.327	0.751	0.249	0.561	0.439	0.561	0.439
Xinyang	0.354	0.646	0.354	0.646	0.276	0.724	0.241	0.759	0.572	0.428	0.218	0.782	0.105	0.895	0.572	0.428	0.572	0.428	0.572	0.428
Zhoukou	0.751	0.249	0.751	0.249	0.639	0.361	0.751	0.249	0.179	0.821	0.673	0.327	0.284	0.716	0.751	0.249	0.561	0.439	0.751	0.249
Zhumadian	0.572	0.428	0.572	0.428	0.572	0.428	0.381	0.619	0.572	0.428	0.459	0.541	0.572	0.428	0.572	0.428	0.381	0.619	0.381	0.619
Jiyuan	0.252	0.748	0.000	1.000	0.354	0.646	0.113	0.887	0.000	1.000	0.218	0.782	0.109	0.891	0.821	0.179	0.821	0.179	0.249	0.751

Table 5.5 Resource condition values in case of variable resource conditions

Technological advance	1999	2000	2001	2002	2003	2004	2005	2006	2007	2008
Zhengzhou	1.506	1.782	1.591	1.428	1.646	1.646	1.428	1.619	1.428	1.782
Kaifeng	1.537	1.595	1.895	1.891	1.623	1.456	1.638	1.891	1.673	1.319
Luoyang	1.467	1.000	1.354	1.000	1.218	1.572	1.572	1.444	1.140	1.603
Pingdingshan	1.494	1.751	1.673	1.891	1.751	1.673	1.751	1.891	1.891	1.891
Anyang	1.506	1.398	1.595	1.891	1.616	1.751	1.179	1.673	2.000	2.000
Hebi	1.887	1.666	1.587	1.778	1.751	1.751	1.673	2.000	2.000	1.428
Xinxiang	1.778	1.398	1.258	1.319	1.179	1.397	1.179	1.669	1.428	1.782
Jiaozuo	1.506	1.619	1.506	1.887	1.428	1.646	1.782	1.895	1.541	1.428
Puyang	2.000	2.000	2.000	2.000	1.646	2.000	2.000	2.000	2.000	1.782
Xuchang	1.000	0.428	0.782	0.922	0.428	1.000	1.000	1.000	0.837	0.782
Luohe	1.619	1.782	1.782	1.817	2.000	2.000	1.533	1.541	1.782	2.000
Sanmenxia	1.113	1.000	1.241	1.467	1.467	1.572	1.572	1.304	1.304	1.304
Nanyang	1.572	1.572	1.000	1.572	1.572	1.572	1.782	1.218	1.241	1.354
Shangqiu	1.534	1.751	1.533	1.456	1.370	1.860	1.673	1.751	1.561	1.561
Xinyang	1.354	1.354	1.276	1.241	1.572	1.218	1.105	1.572	1.572	1.572
Zhoukou	1.751	1.751	1.639	1.751	1.179	1.673	1.284	1.751	1.561	1.751
Zhumadian	1.572	1.572	1.572	1.381	1.572	1.459	1.572	1.572	1.381	1.381
Jiyuan	1.252	1.000	1.354	1.113	1.000	1.218	1.109	1.821	1.821	1.249

Table 5.6 Results of technological advance calculations for Henan cities

Membership	1999	2000	2001	2002	2003	2004	2005	2006	2007	2008
Zhengzhou	1.386	0.539	1.296	1.843	1.116	1.194	0.987	1.062	1.057	1.277
Kaifeng	1.745	0.358	1.252	0.958	1.340	1.081	0.186	0.887	1.963	1.106
Luoyang	1.747	0.736	1.036	1.026	0.903	0.962	1.295	0.952	1.230	1.420
Pingdingshan	1.707	0.967	1.541	0.702	0.989	0.994	0.959	0.577	1.006	0.765
Anyang	2.245	1.797	0.321	1.865	1.233	0.979	1.257	0.773	1.597	1.202
Hebi	0.190	12.516	0.246	0.826	0.996	0.869	0.556	3.406	0.916	1.697
Xinxiang	1.841	0.550	1.177	1.201	1.055	1.082	1.201	1.158	0.798	1.681
Jiaozuo	1.339	2.175	0.434	2.671	0.470	1.099	1.823	0.748	0.837	1.549
Puyang	0.415	0.232	0.253	56.422	1.404	0.242	0.471	0.486	1.949	0.915
Xuchang	1.699	0.795	1.191	0.790	1.648	1.044	0.965	0.867	1.230	1.232
Luohe	0.946	0.736	1.347	0.846	1.088	0.954	0.166	3.025	1.293	1.480
Sanmenxia	1.416	1.475	0.995	0.399	3.722	1.699	0.591	2.612	0.847	1.660
Nanyang	1.669	0.775	1.033	0.886	1.316	1.204	1.017	1.116	0.819	0.945
Shangqiu	1.095	0.177	0.645	1.886	1.658	1.396	0.978	1.390	0.426	1.009
Xinyang	1.327	0.400	2.550	1.218	1.133	0.591	0.834	0.755	1.109	5.240
Zhoukou	1.801	1.092	0.948	0.489	0.875	2.237	0.768	2.156	0.535	0.980
Zhumadian	1.563	0.857	0.733	2.493	0.682	1.091	0.491	2.006	0.684	1.353
Jiyuan	0.393	1.059	2.261	3.076	1.405	1.242	1.654	0.772	1.393	1.035

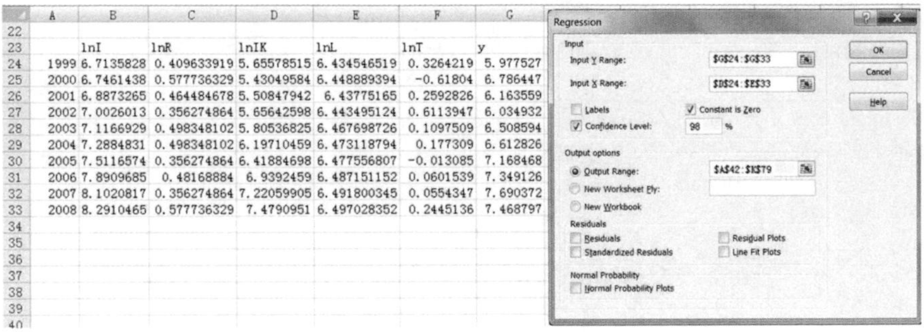

Fig. 5.1 IV regression fit setup for intensive land use in Zhengzhou

5.4 IV Model Fitting

Linear regression was conducted on the IV using logarithms. The results of the fitting process are the elasticity coefficients of capital and labor input of Henan Province and its direct subordinate cities. The data were then processed to obtain the indicator values necessary for IV model fitting: RMB 10,000/km^2, RMB 100 mn, and 10,000 persons for output per unit of land area, total fixed asset investment, and population, respectively. Obtain their logarithms to setup the parameters in the regression screen of the computing tool. For example, the setup for Zhengzhou is shown in Fig. 5.1.

Here,

$$y = \ln I - \ln L - \ln K \tag{5.10}$$

In the regression analysis, L and K represent X_1 and X_2, respectively. Set the constant item to zero. The fitting results are shown in Table 5.7.

As shown here, the multiple R value is 0.9992 and R square is 0.9983, indicating a satisfactory regression fitting result. Based on the fitting process, the elasticity coefficients of capital input and labor input of intensive land use in Zhengzhou are $\alpha = 0.7130$ and $\beta = 0.3609$, respectively. Hence, the IV function for Zhengzhou's intensive land use is

$$I = R \times K^{0.7130} \times L^{0.3609} \times T \tag{5.11}$$

Similarly, the fitting of the IV models of Henan Province and its subordinate cities shows that all of the multiple R values are higher than 0.98 and that the R square values exceed 0.96. See Table 5.8 for the post-regression α and β values of cities across Henan.

Table 5.7 Results of IV model regression analysis for intensive land use in Zhengzhou

Regression statistics	
Multiple R	0.9992
R Square	0.9983
Adjusted R Square	0.8731
Standard Error	0.3107
Observed Value	10

Table 5.8 IV model regression fitting results of cities in Henan province

City	α	β	$\alpha + \beta$
Zhengzhou	0.7130	0.3609	1.0739
Kaifeng	0.4565	0.6272	1.0837
Luoyang	0.5236	0.4473	0.9710
Pingdingshan	0.7293	0.3764	1.1057
Anyang	0.4969	0.5167	1.0136
Hebi	0.3099	0.9512	1.2611
Xinxiang	0.3462	0.6375	0.9837
Jiaozhuo	0.5880	0.5928	1.1808
Puyang	0.4599	0.6608	1.1207
Xuchang	0.5870	0.5313	1.1183
Luohe	0.2762	0.9161	1.1923
Sanmenxia	0.5584	0.4556	1.0139
Nanyang	0.7409	0.2109	0.9518
Shangqiu	0.5027	0.4644	0.9672
Xinyang	0.1381	0.6214	0.7595
Zhoukou	0.6053	0.3533	0.9586
Zhumadian	0.5672	0.3624	0.9296
Jiyuan	0.4320	1.0381	1.4701

5.5 IV Analysis

Total product was adjusted with the resultant IV function model to eliminate the effect of all other factor inputs on the incremental input of any single factor. In other words, adjustments were made to keep all other inputs constant to observe total product changes in response to the increase of a single factor input. With regard to the GDP of Henan Province, two types of adjustments were made to obtain (1) total product change in response to increases of labor input (I_L) per unit of resources, and (2) total product change in response to increases of capital input (I_K) per unit of resources. Then, the trend curves were used to observe total product changes in response to increases of labor input per unit of resources as well as the changes of labor input (L), capital input (K), IVs (I), fitting function-based IVs (I), and adjusted I_L and I_K per unit of total product, as shown in Fig. 5.2.

Obviously, with the effect of technological advances and other external factors, the labor input (L), capital input (K), and IV (I) curves show smooth changes,

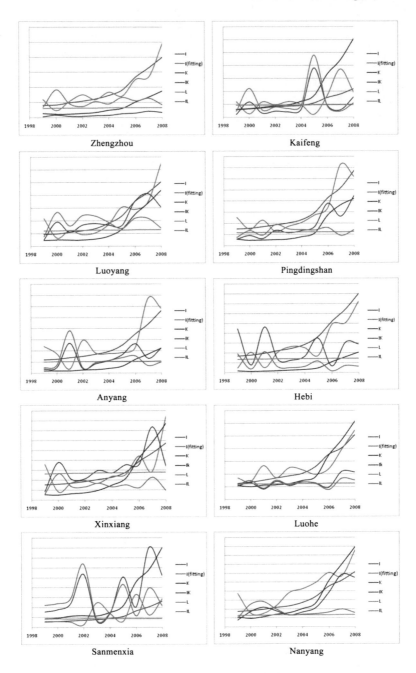

Fig. 5.2 IV curves of land use in Henan province, 1999–2008

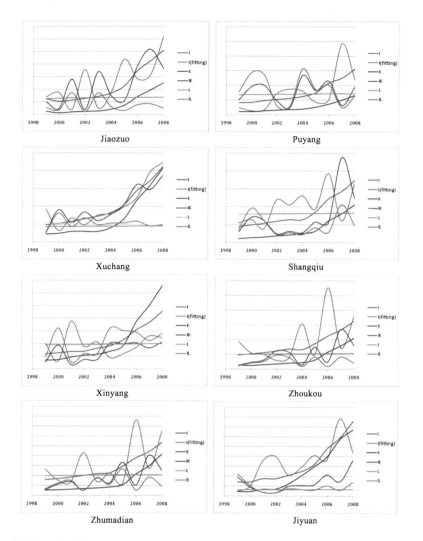

Fig. 5.2 (continued)

whereas the fitting function-based, adjusted I (fitting), I_L, and I_K curves have considerable fluctuations. This is because without the effect of technological advances and other external factors, the total product changes in response to the increase of any single factor, including IV, capital input, and labor input tend to be irregular. It is the constant effect of external factors on IV that causes fluctuations in the labor input (L), capital input (K), and IV (I) curves.

When capital input and labor input increase and technological advances and other inputs change, total product per resource unit increases continuously. Adjustments were made to total product per unit of resources, where capital input and labor input per unit of resources increased while other inputs were kept constant.

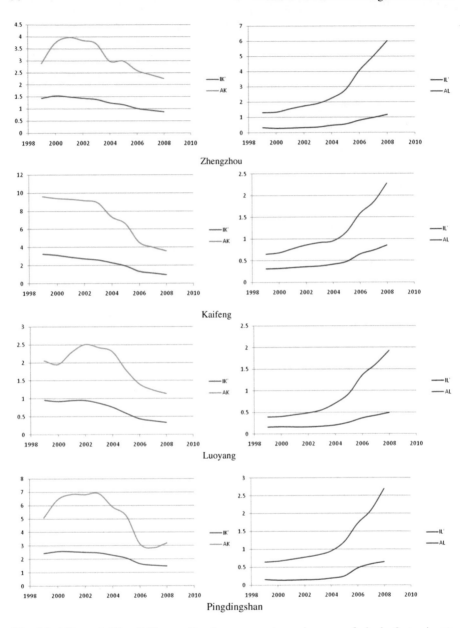

Fig. 5.3 MPs and APs of Henan cities in response to an increase of single factor inputs, 1999–2008

The result of both cases is the increase of total product per unit of resources in similar fluctuating curves. However, this is insufficient to explain the actual situation of intensive land use. As required by the LDR, observations were made on the

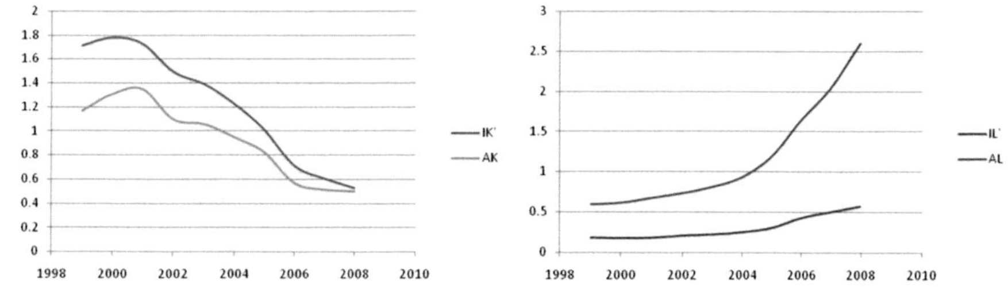

Anyang

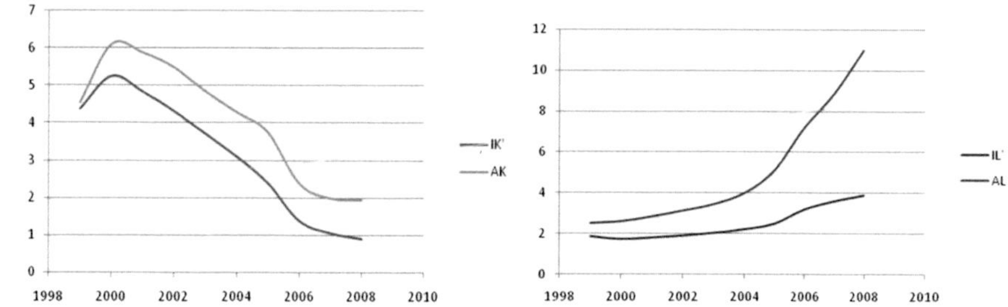

Hebi

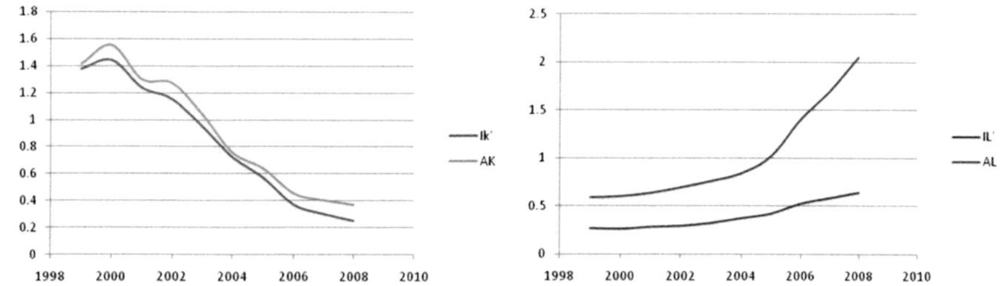

Xinxiang

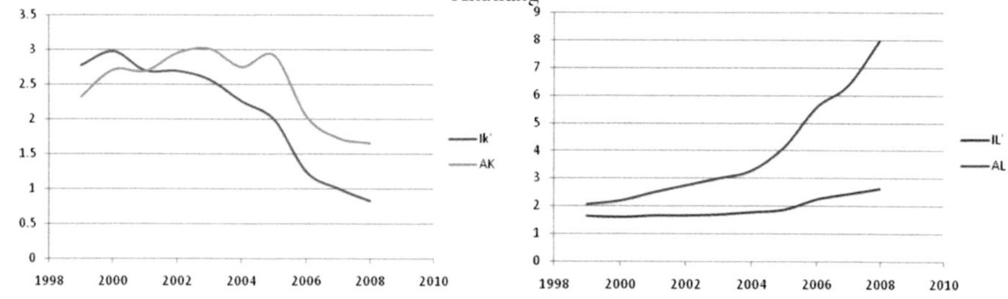

Louhe

Fig. 5.3   (continued)

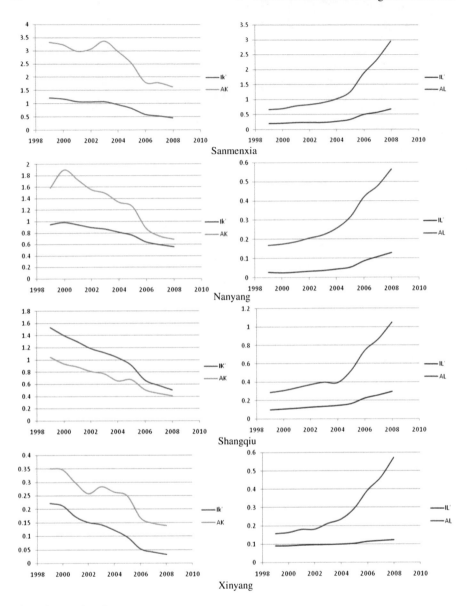

Fig. 5.3 (continued)

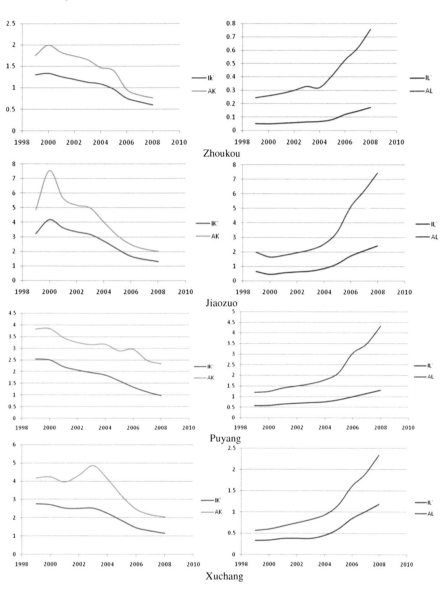

Zhoukou

Jiaozuo

Puyang

Xuchang

Fig. 5.3 (continued)

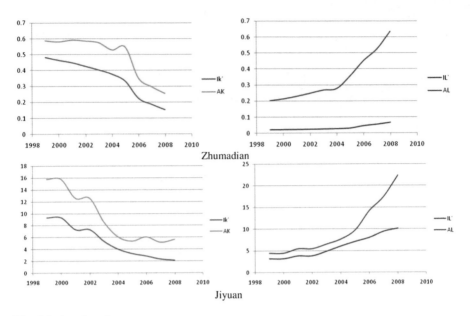

Zhumadian

Jiyuan

Fig. 5.3 (continued)

Fig. 5.4 Intensive land use in Henan intensive land use in Henan

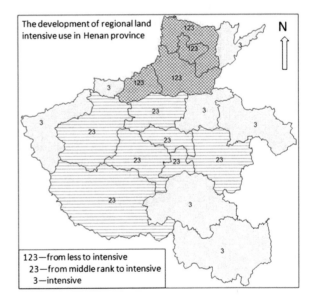

adjusted changes of MPs and APs in response to the increase of a single factor input over the period. Then, judgments were made on the extent of intensive land use in line with the specific development stages involved.

MP of capital input (I'_K), AP of capital input (A_K), MP of labor input (I'_L), and AP of labor input (A_L) per resource unit of the cities in Henan were processed under the assumption that all other inputs are kept constant, as shown in Fig. 5.3.

Here, the IV model fitting covers only a 10-year time span, which is not enough to describe the comprehensive land use situation in the region. Therefore, we used the conditions of the five different stages as a reference and obtained the elasticity coefficient through the fitting process. Then, we made adjustments based on the assumption that technological advances and all other factors remained constant and that only labor or capital inputs changed to determine the development stages of Henan's cities, see Fig. 5.4 for the results. Thereafter, we conducted further analysis on intensive land use in Henan.

The results indicate that based on the condition of capital input, almost every city has a satisfactory rating and is able to achieve intensive land use in later stages. Particularly, Anyang, Hebi, Xinxiang, and Jiaozuo have transitioned from the low-intensity stage to the medium stage and the intensive stage; Zhengzhou, Luoyang, Pingdingshan, Xuchang, Luohe, Nanyang, and Zhoukou have transited from the medium-intensity stage to the intensive stage; and Kaifeng, Puyang, Sanmenxia, Shangqiu, Xinyang, Zhumadian, and Jiyuan are currently in the intensive stage. Based on the condition of labor input, most of the cities in Henan are in the low-intensity stage. Judging by capital input per unit of resources, virtually all cities in Henan are in the intensive development stage, and increasing capital input per unit of resources adequately to the extent where MP is just above zero is a viable method for increasing the level of intensity. Based on labor input per unit of resources, the cities are in the low-intensity stage. This, however, does not indicate insufficient labor input. Instead, it means that current labor input does not match product, resulting in low productivity and low efficiency of labor per unit of resources. In view of the Gini coefficient of Henan's construction land, instead of a rational stage, Henan's construction land is generally in a state of an absolute average. The general results indicate a state of fairly intensive land use instead of fully intensive land use in Henan, and that possibility for improvement still exists in terms of rational use of land resources.

5.6 Summary

Based on the IV theory, this chapter focuses on a case study regression fitting and analysis on the intensive land use model of Henan Province, generating the following findings:

1. The IV model is capable of reflecting the economic development status of a particular region. Based on the condition of capital input, almost every city in Henan has a satisfactory rating and is able to achieve intensive land use in the later stages. Based on the labor input conditions, most of the cities in Henan are

in the low-intensity stage, indicating a less rational use of construction land. The general results indicate that intensive land use in Henan is below a reasonable level, which mirrors the reality of the province. It indicates that even if the simplified IV model is applicable in real-world economic development, it is necessary to escalate from theory to practice; furthermore, the IV model is capable of reflecting the status of intensive economic growth of a particular region.

2. The LDR is incorporated into the IV model.

IVs reflect the changes to the value, benefits, and utility of resources as a result of human activities. In practice, capital input, labor input, and resource/technology conditions are constantly changing. Using the IV function model, it is possible to make adjustments to input changes within a particular region. Simulations of products under the condition of constant inputs generate results that are consistent with the LDR, that is, while technological advances and other factors are kept constant, the increase of a single input results in product changes consistent with the LDR. In other words, the LDR supports the IV theory, which, in turn, validates the LDR.

3. The IV theory is both rooted in and builds on the LDR theory, which is an increasing step function of the former.

Reference

1. Tian, Jun, Pengzhu Zhang, Kanliang Wang, et al. 2004. The integrating model of expert's opinion based on Delphi method. *Systems Engineering Theory and Practice* 1: 57–62.

Chapter 6
Intensive Evaluation of Urban Land Use

Abstract On the basis of IV theory and computing tools discussed in previous chapters, this chapter studies intensive land use evaluation on subordinate districts/counties. China's capital, Beijing, is selected as the target of this case study. The basic functional model of IVs is used in accordance with the dynamic changes of resource conditions, while indicators are selected including GDP, land area, resource condition suitability index, total fixed asset investment, population, technological advance, and land Gini coefficient. Calculation results show Beijing has already been in the intensive land use stage, but is in an excessive use stage spatially. The IV model is proved to be applicable on intensive urban land use evaluation, and IVs can accurately reflect the status of urban land use.

Keywords Case study · Intensive land use evaluation · Urban scale · Beijing

In this chapter, we select China's capital, Beijing, as the target for our IV model fitting to gain insight into the intensive land use status of the subordinate districts/counties. Resource quantity means land areas of the districts/counties. The scope of the resource condition suitability assessment is the entire territory of Beijing. The model selected is a complete basic IV function.

6.1 Data and Pre-processing

All data were abstracted from Beijing-related data, particularly, those for technology development of the city in the China City Statistical Yearbook 1996–2008.[1] Spatial information datasets, including the basic geographical dataset, the IV dataset, and other information datasets were created using the same method as described in Chap. 5. In addition, urban land area was added into the "basic resource information" portion. Resource structure and distribution optimization were obtained through calculation using the land Gini coefficient.

---

[1] http://tongji.cnki.net/kns55/Navi/HomePage.aspx?id=N2010042092&name=YZGCA&floor=1

X. Zheng et al., *Intensive Variable and Its Application*, SpringerBriefs in Geography, DOI: 10.1007/978-3-642-54873-4_6, © The Author(s) 2014

6.2 Evaluation Unit and Indicator Selection

In Chap. 5, we explained that it was possible to conduct a time–spatial analysis on the intensive land use of a particular region. The analysis provides visibility into not only intensive land use of specific jurisdictions of a particular point in time but also dynamic changes in different periods. Focusing on a complete analysis, this chapter provides an integral model to analyze the overall changes to the IV of a particular city. Therefore, Beijing (mainly its urban districts) is selected as the target of our IV study on urban land use.

The indicators in the IV model depend on the actual situation of specific urban districts. Because the focus of this chapter is the IV of urban land use, it is easy to understand that land resource conditions change in line with socioeconomic development goals of different periods. Therefore, our IV study on Beijing's urban land use was conducted in accordance with the dynamic changes to the resource conditions using the basic functional model of IV as its theoretical model, that is

$$I = Q/S \tag{6.1}$$

$$I = RL^{\alpha}K^{\beta}e^{(T+M)} + \delta \tag{6.2}$$

$$Q/S = RL^{\alpha}K^{\beta}e^{(T+M)} + \delta \tag{6.3}$$

Identify the values of relevant indicators in 1996–2008, as shown in Table 6.1.

Specifically, the values of the resource condition suitability assessment indicators were calculated using the same method as that described in Chap. 4. With regard to technological advances, technological input refers to the staffing and funding expenses for R&D, testing, and development activities; technological output refers to research papers published, new product sales of industrial enterprises above the designed scale, Chinese patents granted, and contract values of technology transactions.

Table 6.2 is a summary of indicator values in 1997–2008.

6.3 Indicator Calculation

6.3.1 Resource Condition Suitability Assessment

Resource condition indicators were obtained using the FCE method with the same factor sets and comment sets described in Chap. 5. However, the membership function was developed using a different method. See Table 6.3 for the results of Beijing's primary membership matrix.

Table 6.1 Urban Land IV indicators

Indicator	Q	S	R	K	L	T	M
Meaning	GDP (RMB 10 K)	Land area (km²)	Resource condition suitability index	Total fixed asset investment (RMB 100 mn)	Population (10 k)	Technological advance	Land Gini coefficient

Table 6.2 Results of IV data collection for Beijing's urban land

Year	GDP (RMB 100 mn)	Land area (km²)		Year end population (10 k)	Fixed asset investment (RMB 100 mn)
		Area	District		
1997	1,810.09	16,410.54	2,642.45	1,216.70	582.08
1998	2,011.31	16,410.54	2,672.62	1,223.39	682.92
1999	2,174.46	16,410.54	2,699.78	1,249.90	651.40
2000	2,478.76	16,410.54	2,731.12	1,107.53	670.57
2001	2,845.65	16,410.54	2,868.80	1,122.30	1,417.07
2002	3,212.71	16,410.54	3,086.47	1,136.30	1,814.30
2003	3,663.10	16,410.54	3,024.54	1,148.82	2,157.10
2004	4,283.31	16,410.54	3,197.23	1,162.89	2,528.30
2005	6,886.31	16,410.54	3,230.23	1,180.70	2,827.20
2006	7,870.28	16,410.54	3,272.64	1,197.60	3,371.50
2007	9,353.32	16,410.54	3,325.57	1,213.26	3,966.57
2008	10,488.05	16,410.54	3,338.16	1,299.85	3,848.55

Table 6.3 Beijing's resource condition FCE primary membership calculation results

Membership	1999	2000		2001		2002		2003		2004		2005		2006		2007		2008	
u11	1	0	0	1	0	1	0	1	0	1	0	1	0	1	1	0	1	0	1	0
u12	0	1	0	1	0	1	0	1	0	1	1	0	1	0	1	0	1	0	1	0
u13	0	1	0	1	0	1	0	1	0	1	0	1	1	0	1	0	1	0	1	0
u21	0	1	0	0	1	0	1	1	0	0	1	1	0	1	0	1	0	1	0	0
u22	1	1	0	0	1	0	1	1	0	0	1	1	0	1	0	1	0	1	0	1
u23	0	0	1	1	0	0	1	1	0	0	1	1	0	1	0	1	0	1	0	0
u31	0	0	1	0	1	0	1	1	0	1	0	1	0	1	0	1	0	1	0	0
u32	1	0	1	0	1	0	1	0	1	0	1	1	0	1	0	1	0	1	0	1

Weights of the factors were re-identified through two rounds of scoring by 13 experts using the Delphi method in accordance with the specific requirement of urban development. The resultant weights for the primary factors were as follows:

$$A_1 = (0.377, 0.304, 0.319)$$
$$A_2 = (0.371, 0.336, 0.293)$$
$$A_3 = (0.477, 0.523)$$

Table 6.4 Beijing's FCE primary judgment results

Membership	1999		2000		2001		2002		2003	
U1	0.419	0.581	0.000	1.000	0.000	1.000	0.000	1.000	0.000	1.000
U2	0.682	0.318	0.779	0.221	0.221	0.779	0.000	1.000	1.000	0.000
U3	0.481	0.519	0.000	1.000	0.000	1.000	0.000	1.000	0.481	0.519
Membership	2004		2005		2006		2007		2008	
U1	0.254	0.746	0.581	0.419	1.000	0.000	1.000	0.000	1.000	0.000
U2	0.000	1.000	1.000	0.000	1.000	0.000	1.000	0.000	1.000	0.000
U3	0.481	0.519	1.000	0.000	1.000	0.000	1.000	0.000	1.000	0.000

Table 6.5 Beijing's FCE membership calculation results

Year	Membership		Year	Membership	
1999	0.526	0.474	2004	0.214	0.786
2000	0.276	0.724	2005	0.821	0.179
2001	0.078	0.922	2006	1.000	0.000
2002	0.000	1.000	2007	1.000	0.000
2003	0.459	0.541	2008	1.000	0.000

The weights for the advanced factors were

$$A = (0.428, 0.361, 0.211)$$

Calculate the fuzzy compound matrix. First, conduct a primary judgment. See Table 6.4 for the results.

Then, conduct secondary judgment. See Table 6.5 for the results.

If resource conditions are variable, their value in the IV model is set between 1 and 2 to reflect the features of resource condition changes and to differentiate the case from that of constant resource conditions. Based on the results of this FCE, the value of the resource conditions in the variable case is obtained by adding 1 to the "suitable" membership of the comment set. See Table 6.6 for the results.

6.3.2 Technological Advances

In accordance with the requirements for the calculations, relevant data were collected and processed, as shown in Table 6.7.

Beijing's technological advances of 1999–2008 were then calculated using the MIs (see Table 6.8).

Table 6.6 Resource condition values in case of variable resource conditions

Year	RC value	Year	RC value
1999	1.526	2004	1.214
2000	1.276	2005	1.821
2001	1.078	2006	2.000
2002	1.000	2007	2.000
2003	1.459	2008	2.000

Table 6.7 Results of technological advance calculations of Beijing

Year	Essays published	New product sales of EADS (RMB 100 mn)	Chinese patents granted	Contract value of tech transfer (RMB 10 k)	R&D expense (RMB 100 mn)	R&D and test staffing (person-years)
1998	6,971	125	3,800	815,591	103.3	87,004
1999	11,541	256	5,829	921,889	121.6	84,731
2000	12,536	515	5,905	1,402,871	155.7	98,753
2001	14,507	571	6,246	1,910,065	171.2	95,255
2002	17,586	640	6,345	2,211,738	219.5	114,919
2003	21,393	302	8,248	2,653,574	256.3	109,947
2004	23,533	510	9,005	4,249,975	317.3	151,542
2005	34,674	342	10,100	4,895,922	382.1	171,045
2006	36,578	1,229	11,238	6,973,256	433.0	168,398
2007	41,162	2,429	14,954	8,825,603	505.4	187,578
2008	48,076	2,471	17,747	10,272,173	550.3	189,551

Table 6.8 Results of Beijing's technological advance calculations

Year	Technological advance	Year	Technological advance
1999	1.421	2004	1.039
2000	1.169	2005	0.853
2001	1.165	2006	1.843
2002	0.892	2007	1.308
2003	0.741	2008	1.047

6.3.3 Other Relevant Indicators

In the intensive land use model of Beijing, the IV is GDP per land area unit, or RMB 10 k/km^2.

Resource structure and distribution optimization were calculated using Gini coefficients. Table 6.9 shows Beijing's urban construction land Gini coefficients calculated for 1999–2008. See Fig. 6.1 for the development trend.

Table 6.9 Beijing's urban construction land Gini coefficients in 1999–2008

Year	1999	2000	2001	2002	2003	2004	2005	2006	2007	2008
LGC	0.452	0.449	0.441	0.428	0.432	0.423	0.420	0.418	0.415	0.414

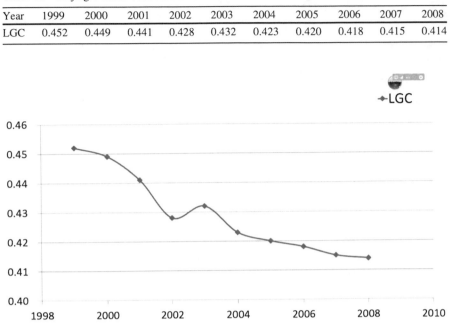

Fig. 6.1 Development trend of Beijing's urban construction land Gini coefficient in 1999–2008

Figure 6.1 shows that Beijing's urban construction land Gini coefficients were above 0.4 in 1999–2008, indicating that land use structures were totally unbalanced or inadequate.

Other indicators were used directly for IV model fitting.

6.4 IV Model Fitting

A linear regression was conducted on the IV using logarithms. The results of the fitting process were the elasticity coefficients of capital input and labor input of Beijing. The data were then processed to obtain the indicator values necessary for IV model fitting, and the logarithms for the regression calculation were conducted in the same manner as that described in Chap. 5. To setup the parameters in the regression screen of the computing tool, apply the data to the formula for fitting. See Table 6.10 for the results.

As shown here, multiple R is 0.9995, and R square is 0.9991, indicating a satisfactory regression fitting result. Based on the fitting process, the elasticity coefficients of capital input and labor input of intensive land use in Beijing are $\alpha = 0.2705$ and $\beta = 0.5516$, respectively. Hence, the IV function for Beijing's intensive land use is as follows:

Table 6.10 Results of IV model regression analysis for intensive land use in Beijing

Regression statistics	
Multiple R	0.9995
R square	0.9991
Adjusted R square	0.8740
Standard error	0.2518
Observed value	10

$$I = RK^{0.5516}L^{0.2705}e^{(T+M)} \tag{6.4}$$

In comparison, we also conducted a calculation using the method described in Chap. 5 and obtained $\alpha = 0.5201$, $\beta = 0.5073$, and the function is

$$I = R \times K^{0.5201} \times L^{0.5073} \times T \tag{6.5}$$

Through a comparison, we found that the basic IV function model was better capable of reflecting the effects of technological advances and management improvements. In other words, given a constant capital contribution, incremental technology and structural factors have resulted in a significant reduction of the contribution of labor. The development of human society is a process of continued increases in capital inputs. However, the development of technologies is one with a diminishing labor input along with technological advances and management improvements, which is the basic law of socioeconomic development of human society.

6.5 IV Analysis

The regression results show that the elasticity coefficient of capital input is higher than that of labor input. Based on the IV model, adjustments were made to GDP and technological advances to observe total product changes in response to the increase of a single factor input, whereas all other factor inputs were kept constant. Figure 6.2 illustrates the IV (I), capital input (K), and labor input (L) before and after the adjustment. Only the IV of capital input (IK) and labor input (IL) show notable changes in their development trend after the adjustment.

The figure shows that the biggest change to the curves after the adjustment is that they fluctuate more, that is, the change of a single factor has more distinct non-linear properties. A smoother curve results, owing to interaction between different factors. Regardless of the changes, there is a general upward trend.

As shown in Fig. 6.2, both capital and labor inputs increase over time. Given the effect of resource conditions and technology activities, the IVs increase each year. After adjusting the IV model and the effects of technology activities and incremental capital/labor inputs, the IVs of any single input increase despite the fluctuations.

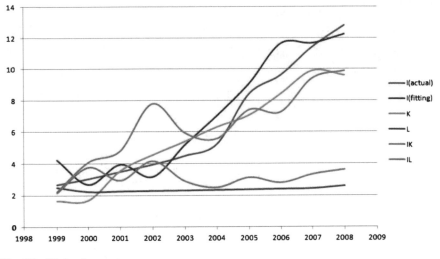

Fig. 6.2 IV development *curve*

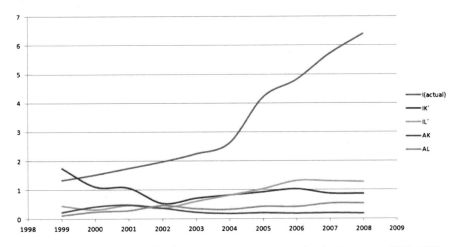

Fig. 6.3 MPs and APs of Beijing in response to an increase of a single input in 1999–2008

Figure 6.3 illustrates the MP (I'_K) and AP (A_K) of capital input per unit of resources and the MP (I'_L) and AP (A_L) of labor input per unit of resources of Beijing, provided that all other inputs are kept constant.

Based on IV theory, we observed the post-adjustment MP and AP changes in response to the increase in a single input over time. Then, we determined the level of intensive development in line with the production stages involved. The result shows that considering capital input alone, Beijing's land use is in Stage III, where the MP of incremental capital input is just above zero and the AP declines. Its IV is near the maximum value or approaching the most intensive stage. By contrast, when considering labor input alone, Beijing's land use is in Stage I, the low-intensity stage. Obviously, labor productivity is low, and both MP and AP of incremental labor input are still rising. Thus, there is still room for further development, and incremental labor input remains the primary approach for Beijing's socioeconomic growth. In view of Beijing's land Gini coefficient, Beijing's urban land use structure seems substantially unbalanced. In general, Beijing's land use is in the intensive stage, whereas its spatial structure is in the over-intensive use stage. Therefore, it is necessary to improve productivity and to ensure rational planning and intensive use of urban land.

6.6 Summary

Based on the basic IV function, this chapter focuses on a case study regression fitting and analysis on the intensive land use model of Beijing. Our findings are as follows:

1. The IV model is applicable to intensive urban land use evaluation.

Using a city's available land resources, economy, population, and other related data, we conducted IV model fitting and obtained a captial input elasticity coefficient $\alpha = 0.5516$ and labor input elasticity coefficient $\beta = 0.2705$. Then, we drew the IV, MP, and AP curves based on the assumption that all factors other than capital and labor inputs remained constant. Using these curves as a reference, we determined what stage of intensive development the city was in and obtained results that were consistent with the actual situation of Beijing. This proves that the IV model is applicable to intensive urban land use evaluation. Therefore, we have escalated the IV from the theoretical layer to the practical layer.

2. The economic development of the city is consistent with IV theory.

IVs reflect the changes to the value or benefits of land resources as a result of human activities. Using the IV function model, it is possible to make adjustments to input changes of the city. The results of the product simulations under constant inputs conditions are consistent with the actual situation of Beijing, indicating that the city is still in the intensive development stage.

3. IVs can accurately reflect the status of land use.

The calculation results indicate that in the model shown previously, we can use different data for the IV calculation to obtain data regarding specific time points, spatial distribution, and land use structures. These results can be analyzed separately or in combinations to provide decision makers with better references.

Chapter 7
Conclusion and Outlook

Abstract IV is variable that reflect the increase of value, benefits, and utility of natural resources as a result of human activities. On the basis of LDR theory and a new discovery on economic development principles, the IV theory was developed along with the theoretical model, basic function, and curves to gain insight on the effect of human activities on natural resources. We also described our approaches and results of our estimation on resource conditions, technological advances, resource structures, and distribution optimization to ensure the applicability of our research findings in practice. A GIS-based computing tool has been developed for an automated calculation. Case studies on regional and urban scales have been conducted respectively on Henan province and Beijing. IV theory offers a unique framework of theories and methodologies for the evaluation of natural resource use, thus ending the lack of generally acceptable theories and methodologies. It will have a profound effect on economic development theories from a more comprehensive and macroscopic standpoint, and will provide a better picture of resource use from time-spatial perspectives. However, more effort must to be made to further improve the IV theory system.

Keywords Intensive Variable theory · LDR theory · Intensive evaluation · GIS-based computing tools · Case study

7.1 Conclusion

IVs are variables that reflect the increase of value, benefits, and utility of natural resources as a result of human activities. The development of IV theory was inspired by the LDR. Traditionally, intensification was believed to be a state of the LDR. However, based on a study of the history of human society's economic activities, we discover that intensive development is an increasing function of sustainability, and that the state of diminishing returns is only one of the states observed during the course of intensive development. This novel discovery

X. Zheng et al., *Intensive Variable and Its Application*, SpringerBriefs in Geography, DOI: 10.1007/978-3-642-54873-4_7, © The Author(s) 2014

changes the basic understanding regarding the LDR over the past two centuries, as well as the basic concept of intensification. It enables a review of the LDR from a higher viewpoint [1, 2]. The LDR is essentially a function used to determine the turning point in IV theory.

In reality, along with technological advances, productivity and returns might increase. Thus, this theory breaks the LDR under ideal conditions but does not deny its existence. In general, economic development is a process that evolves from the extensive stage to the intensive stage, and from lower models to higher models. It is therefore necessary to drive the transformation of economic growth toward an intensive model. Specifically, actions should be taken to maximize intensive growth and to drive the economy from the low-intensity stage toward the medium- and high-intensity stages. Inspired by intensive economic development ideas, the IV theory is developed along with the theoretical model, basic function, and curves to gain insight on the effect of human activities on natural resources. These findings can help expand economic development theories from a more comprehensive and macroscopic standpoint. In addition, this approach can provide a better picture of resource use from time-spatial perspectives.

Using the same method to estimate the production function, we described our estimation of resource conditions, technological advances, resource structures, and distribution optimization to ensure the applicability of our research findings in practice. We proposed a number of methods for parameter estimation, including FCE, the MI method, the Gini coefficient approach to estimate resource condition coefficients, technological advances, resource structures, and distribution optimization extent, respectively.

In order to simplify complex calculations, we developed a GIS-based computing tool, which enabled evaluation unit selection, parameter calculation, model fitting, and better visibility of the entire calculation process and its results.

Based on the IV theory, this book evaluates intensive land use in Henan Province and Beijing using both simplified and standard IV functions. First, in the case of Henan Province, we used the simplified model to calculate the IVs and analyze the identities of intensive land use in the prefectural cities in 1999–2008. Then, for Beijing, we used the standard model to calculate the IVs of the city for 1999–2008. Next, we roughly compared the effects of different function models on the calculation structure and concluded that the standard model was able to provide better explanations. Using the models obtained from the simulation process, we adjusted the total product and obtained the IVs resulting from the increase of a single input while all other factors remained constant. Thereafter, we drew the post-adjustment IV, MP, and AP curves for the evaluation period. Referencing the IV curve, we conducted an overall assessment on intensive land use in Henan and Beijing and drew the following practical conclusions: In Henan Province, most land is currently in the transition from the medium-intensity stage to the intensive stage and has yet to progress into the fully intensive stage. This shows a mean distribution from a spatial perspective that differs from the rational state. By contrast, Beijing has already been in the intensive land use stage. However, spatially, it is in an excessive use stage.

The IV theory attempts to provide a macroscopic explanation on the content of intensive use, that is, a phenomenon resulting from human exploitation of natural resources. The IV theory has extended the content of intensive use research to a wider time-spatial scope, enabling new developments over locations and periods. The IV model is a quantitative form of IV theory. The integration of the IV theory with relevant computer models and tools allows an in-depth analysis of intensive resource use from different time-spatial perspectives. Being feasible not only in theory but also in practice, it helps improve the visibility of the theoretical calculation process and is proved to be an intensive use evaluation approach applicable on a wider scope.

In summary, in this book, we used IVs as our theoretical basis and the time-space correction and IV model as basic tools to integrate the IV curve with the LDR. With the support of GIS, we attempted to combine static and dynamic data and to integrate mathematic calculations with spatial optimization. The result is a unique framework of theories and methodologies for intensive land use evaluation. Then, we extended the theories into the evaluation of natural resource use, thus ending the lack of generally acceptable theories and methodologies. We believe that our work will have a profound effect on both theoretical research and practice in this particular field. However, more effort must be made to further improve the IV theory system.

7.2 Outlook

The IV theory is based on the assumption that natural resources are available and can generate additional value, benefits, and utility as a result of human activities. According to its original definition, IV theory is applicable to the evaluation of intensive land use as well as other natural resources. The variable in the IV model may stand for different indicators, depending on the specific resource type and usage. In other words, the indicators for resource utilization should reflect the backgrounds and purposes of specific studies. For example, a research model for the evaluation of intensive exploitation of a particular type of mineral resource can use the resource's generated value to indicate its benefit, the input amount to indicate resource quantity, the quality assessment result to indicate resource conditions, the amount of money used to develop and exploit the resources to indicate capital input, the number of workers employed to indicate labor input, and relevant technical indicators to indicate technological advances (T). See the China Statistical Yearbook on Science and Technology for more details. Indicator values are collected for IV model fitting. In this study, we used the elasticity coefficients to adjust the IV to fit the ideal condition scenario. Next, we observed the development trends of the IV, analyzed the evolement of its structural from a spatial perspective, and made a general judgment regarding intensive resource utilization. Due to the restriction of the book length, we did not discuss specific case studies in depth.

Even with regard to land use evaluation, the application scope of IV theory can be further expanded to include the evaluation of other land resource types, such as farming land [3, 4], land owned by villages [5, 6] or development zones [7, 8], and land designated for education [9] purposes. Indicators for the IV models should be identified in accordance with the identities of different land types. For example, for the evaluation of intensive use of farming land, indicators relevant to its quality should be selected for resource condition suitability assessment. Furthermore, benefits of farming land, resource quantity, and technological advances can be indicated using grain yields, land area, and chemical fertilizer productivity and other technical inputs (e.g., hi-tech planting), respectively. For the evaluation of intensive use of land owned by villages, resource conditions, resource conditions for suitability assessment, benefits of land, and technological advance can be respectively specified using the land area of specific villages, living and production conditions, gross product, and hi-tech planting and other relevant indicators. For land with designated functions, indicators can be identified in line with specific uses for IV model fitting. For example, for land designated for education purposes, the benefits can be indicated using the number of people that have graduated or become employed. Furthermore, capital input can be indicated using education expense, labor input using the number of teachers, resource conditions (for suitability assessment) using educational environment and equipment, and technological advance using essays published, patents granted, lab equipment, and infrastructure. If the indicators in the IV model selected are consistent with the meanings and can justify their assumptions, the model can be used to assess intensive use of the evaluation target.

In this book, we used a land Gini coefficient to analyze the rationality of the land use structure. For the same purpose, a number of other methods can be considered. For example, the landscape shape index (LSI) [10] can describe the rationality of specific spatial structures by incorporating the results of spatial tests into the IV model for a time-spatial intensive use analysis.

Currently, IV model fitting is conducted using two established computing tools, that is, ArcGIS and Excel. Based on the fitting process, we must consider secondary development and design tools for independent IV model fitting and automated drawing of the IV curves of the evaluation targets so that the system can automatically determine the extent of intensification of the evaluation target [11, 12].

These issues remain topics for future in-depth studies, and further issues not discussed here may exist. We would be more than happy to see our readers join the discussion.

Acknowledgments This study was supported by the following grant: the National Public Benefit (Land) Research Foundation of China (No. 201111014); the National Natural Science Foundation of China under Grant (No. 40571119). We would like to thank Lina Lv and Dongsheng Hong for valuable comments and suggestions in discussions and Dongsheng Hong's work for the development of computational tools. We would also like to express our gratitude to editors of editage for their helps to this manuscript.

References

1. Hua, Qing. 1986. Discussion on motion law system of land return. *Journal of Agrotechnical Economics* 6: 29.
2. Gu, Liubao, and Mingqian, Zhang. 2001. The empirical analysis in the "poverty trap" theory of CES model of economic growth. *Statistical Research* 12: 11–14.
3. Chen, Yuqi, and Xiubin, Li. 2009. Structural change of agricultural land use intensity and its regional disparity in China. *ACTA Geographic Asinica* 64(4): 469–478.
4. Zhang, Haiying, and Chun, Fu. 2010. Quantitative analysis of economic growth and cultivated land intensive use of Jiangxi province. *Resources and Environment in the Yangtze Basin* 19(10): 1159–1163.
5. Zhao, Xiaomin, and Jianyu, Deng. 2012. Study on evaluation of rural residential land saving and intensive use. *Resources and Industries* 14(002): 54–59.
6. Zhao, Ruoxi, Changchun, Feng, and Xiaolong, Liu. 2012. Factors analysis of rural residential land saving and intensive use. *Rural Economy* 2: 38–42.
7. Dong, Guanglong, Xinqi, Zheng, Hang, Su, et al. 2012. SEM validation studies: evaluation indicator system on land intensive utilization in development zone. *China Land Science* 26(9): 35–40.
8. Zhang, Yong, and Jingsong, Zhang. 2012. The analysis and evaluation of land intensive use of development zone. *Agricultural Science and Technology and Information* 14: 9–11.
9. Tan, Shukui, and Man, Zhou. 2012. Response of Universities in Wuhan to the intensive land use policy. *Resources Science* 34(1): 143–150.
10. Zheng, Xinqi, and Meichen, Fu. 2010. *Landscape pattern spatial analysis technology and its application*. moscow: Science Press.
11. Shephard, Ronald W., and R. Fire. 1974. The law of diminishing returns. *Zeitschrift für Nationalökonomie* 34: 69–90.
12. Charles, Lees. 2004. Environmental policy: the law of diminishing returns? *PSA Annual Conference*. UK: University of Lincoln. 4 May 2008.

About the Author

Professor Zheng's major is geography and geographic information science. He has long been engaged in teaching and research on land information technology and its applications, land intensive use technology, and engineering. He integrated both researches on optimization allocation and intensive use of urban land, and pioneered the in-depth exploration on the linkage between these two research facets. His published book entitled. "Optimization of urban land allocation and evaluation of intensive use", which gives an account of this novel research method, has gained him international reputation. He is responsible for more than 40 national research programs and has published 11 monographs, over 200 papers with 40 indexed by SCI/EI, and owned 5 invention patents. He is the recipient of numerous awards and honors including 8 provincial and ministerial level scientific and technological awards, and "Scientific Chinese of 2010". As the executive director of National Society and expert of the commission, he also serves in the assessment panel for national and provincial-level awards.

X. Zheng et al., *Intensive Variable and Its Application*, SpringerBriefs in Geography,   79
DOI: 10.1007/978-3-642-54873-4, © The Author(s) 2014